50 Calculation Shortcuts ALL Students Must Know

© 2012 MatheMagics Asia Private Limited

All rights reserved.

Published in the USA by MatheMagics Asia Private Limited
Email: contact@mathemagicsasia.com

ISBN – 13: 978-1-4701-5073-0

Author : Kamal

First Edition

All rights are reserved.
No part of this publication may be reproduced, stored in a retrieval system or transmitted in any form or by any means, elect onic, mechanical, photocopying, recording or otherwise, without prior permission from the publisher

CONTENTS

Page No.

Introduction

How to use this book

01: Adding 9 to a number	10
02: Adding 99, 999 to the numbers	13
03: Subtracting a number from a sequence of 9	16
04: Subtracting a number from a multiple of 10	19
05: Split the adder	22
06: Add in parts	25
07: Add more or less than needed	30
08: Subtract more than necessary	33
09: Adding a series of consecutive numbers from 1 to 'n'	36
10: Sum of first 'n' even numbers	39
11: Sum of first 'n' odd numbers	43
12: Multiplying a number by a multiple of 10	47
13: Multiplying decimal numbers by a multiple of 10 (e.g. 10,100,1000 etc)	49
14: Dividing by 10 or higher multiples of 10	52
15: Multiplying numbers between 10 and 19	57
16: Subtract in Steps	61
17: Subtract by simplifying	64
18: Multiplying a number by 5	67
19: Dividing a number by 5	70
20: Subtract in groups	73
21: Multiplying a number by 9	77

CONTENTS

	Page No.
22: Multiplying a number by 11	81
23: Multiplying a number by 25	89
24: Squaring numbers that end in 5	92
25: Multiplying two number with same first digit and last digits that add to 10	95
26: Squaring numbers near 50	99
27: Squaring numbers that end in '0'	102
28: Squaring numbers that end in '1'	106
29: Multiplying numbers that are equidistant from a multiple of 10	109
30: Adding a series of consecutive numbers from 'a' to 'b'	115
31: Adding a series of numbers from 'a' to 'b'	120
32: Multiplying numbers with same initial digit and other digits of both numbers add up to 100	127
33: Dividing numbers by 9	132
34: Multiplying numbers close to and below 100	137
35: Multiplying numbers close to and above 100	142
36: Multiplying numbers close to and below 1000	148
37: Multiplying numbers close to and above 1000	154
38: Squaring numbers near 100	160
39: Squaring numbers near 1000	166
40: Squaring numbers near 200	174
41: Squaring numbers near 500	180
42: Squaring numbers that end in 25	186
43: Squaring numbers in the 300s	190

CONTENTS

Page No.

44: Squaring numbers in the 400s 194

45: Sum of squares of first 'n' numbers 198

46: Sum of squares of 'n' consecutive numbers 202

47: Sum of cubes of 'n' consecutive numbers 205

48: Adding Time 208

49: Checking Answers: Digit Sums 212

50: Cubing 2-digit Numbers 217

Introduction

What is the importance of learning methods to do mathematical calculations in your head?

Firstly, being able to calculate mentally helps in developing an impeccable understanding of numbers which is the first step towards mastering numbers and being truly numerically literate; a skill in short supply and always more in demand.

Secondly and most relevant one is that students world over are faced with an extremely competitive environment where seconds matter, and not all of these situations allow the use of calculating devices (aka calculators). These methods lead to a distinct advantage in competitive and school examinations.

Thirdly, our schooling system generally leaves us poorly equipped to handle the day to day numerical situations that we encounter, like budgeting, tipping, managing interest rates, credit card bills, mortgages etc. Not only do we remain unexposed to these concepts in school, we grow up with a lack of very basic knowledge of numbers that makes our eyes glaze over in every situation that requires us to add two numbers. The methods in this book will not only create a feeling of understanding but also confidence of not running away from numerical situations

Fourthly, despite decades of growth and advancement in science and technology, the human brain still is your most trusted and handy tool that is probably getting rusted with lack of use. These methods help you use your brain and give it a much needed and timely exercise that it deserves.

With these thoughts, we have written this book as an introduction to the methods of mental calculations. This books contains the most useful ones that would help most of the students and is the first in a series of books on mental methods for calculations.

Galileo said, the universe is written in the language of mathematics. Let this be the first step towards understanding the alphabet of that language. Jump into it. Have fun.

While we have been meticulous in our checking, there still might be some that have slipped past us. If you find any such error, please email it to us at:
mentalmaths@mathemagicsasia.com

All errata will be maintained on the website at:
www.mathemagicsasia.com/50methods.html

Kamal

How to use this book

'50 Calculation Shortcuts' has been designed as a workbook to enable the learner to study the concept and practice it immediately to ensure that the method is understood and also used. This ensures that the method 'sticks' in the minds of the reader.

The recommended way to use this book is as follows:

1. Read through the concept of the shortcut. Read it again if you feel you have not grasped it.
2. Go through the solved examples to see how the method has to be applied.
3. Do the exercises in the book to apply the shortcut.
4. We strongly suggest that when you start reading through a shortcut, you finish it completely, all the way from the rule to the exercises. Leaving the exercises to a later time will make the learning incoherent.

Further support:

1. On our website, you can sign up for a practice account by which you will get timed tests for practicing the shortcuts.
2. If you have any doubts, send an email to us and we will be happy to help you with your query.

50 Calculation Shortcuts

Shortcut 1: Adding 9 to a number

Rule
Adding 9 to a number is same as adding 10 and then subtracting 1 from the result. Because 9 + 1 = 10, it implies 9 = 10 − 1.

Illustration

Add 17 + 9
9 = 10 − 1
17 + 9 = 17 + 10 − 1
= 27 − 1
= 26

Examples:

23 + 9
= 23 + 10 − 1
= 33 − 1
= 32

36 + 9
= 36 + 10 − 1
= 46 − 1
= 45

77 + 9
= 77 + 10 − 1
= 87 − 1
= 86

95 + 9
= 95 + 10 − 1
= 105 − 1
= 104

142 + 9
= 142 + 10 − 1
= 152 − 1
= 151

325 + 9
= 325 + 10 − 1
= 335 − 1
= 334

779 + 9
= 779 + 10 − 1
= 789 − 1
= 788

1249 + 9
= 1249 + 10 − 1
= 1259 − 1
= 1258

Exercise:

1)   18 + 9
= 18 + 10 − 1
= ☐ − 1
= ☐

2)   46 + 9
= 46 + 10 − 1
= ☐ − 1
= ☐

3)   64 + 9
= ☐ + ☐ − 1
= ☐ − 1
= ☐

4)   83 + 9
= ☐ + ☐ − 1
= ☐ − 1
= ☐

5)   88 + 9
= ☐ + ☐ − 1
= ☐ − 1
= ☐

6)   126   +   9 =
=   ☐   +   ☐   − 1
=   ☐   −   1
=   ☐

---

7)   288   +   9
(add 10 directly to the first number and subtract 1 from it.)
=   ☐   −   1
=   ☐

---

8)   1044   +   9
=   ☐   −   1
=   ☐

---

9)   2129   +   9
=   ☐   −   1
=   ☐

---

10)   100234   +   9
=   ☐   −   1
=   ☐

Shortcut 2: Adding 99, 999 to the numbers

Rule
This is a general case of shortcut 1.
You should see that 99=100−1, 999=1000−1, 9999=10000−1 and so on. To add 99 to a number add 100 to the number and subtract 1 from the result.
Similarly to add 999 to a number add 1000 to the number and subtract 1 from the result.

Illustration

Add 866 + 99
Add 100 and subtract 1
= 866 + 100 − 1
= 966 − 1
= 965

Examples:

468 + 99
= 468 + 100 − 1
= 568 − 1
= 567

276 + 99
276 + 100 − 1
= 376 − 1
= 375

553 + 99
= 553 + 100 − 1
= 653 − 1
= 652

501 + 99
= 501 + 100 − 1
= 601 − 1
= 600

1093 + 999
= 1093 + 1000 − 1
= 2093 − 1
= 2092

3248 + 999
= 3248 + 1000 − 1
= 4248 − 1
= 4247

2465 + 999
= 2465 + 1000 − 1
= 3465 − 1
= 3464

1526 + 999
= 1526 + 1000 − 1
= 2526 − 1
= 2525

2341 + 999
= 2341 + 1000 − 1
= 3341 − 1
= 3340

Exercise:

1)   652   +   99
=   652   +   100   −   1
=   ☐   −   1
=   ☐

2)   813   +   99
=   813   +   100   −   1
=   ☐   −   1
=   ☐

3)   133   +   99
=   133   +   100   −   1
=   ☐   −   1
=   ☐

4)   387   +   99
=   387   +   ☐   −   ☐
=   ☐   −   ☐
=   ☐

5)   391   +   99
=   391   +   ☐   −   ☐
=   ☐   −   ☐
=   ☐

6) 1184 + 999
= ☐ + ☐ − ☐
= ☐ − ☐
= ☐

7) 2893 + 999
= ☐ + ☐ − ☐
= ☐ − ☐
= ☐

8) 5254 + 999
= ☐ + ☐ − ☐
= ☐ − ☐
= ☐

9) 8632 + 999
= ☐ + ☐ − ☐
= ☐ − ☐
= ☐

10) 9936 + 999
= ☐ + ☐ − ☐
= ☐ − ☐
= ☐

Shortcut 3: Subtracting a number from a sequence of 9

Rule
Example is subtracting a number from 9, 99, 999 or other larger numbers consisting of only 9s.
This shortcut for subtraction works from <u>left to right.</u>
Look at the number being subtracted and starting from left subtract each digit from 9.

Illustration

99 – 18 =
We subtract the left digit 1 from 9 to get 9 – 1 = 8.
Then we subtract the 8 from 9 to get 9 – 8 = 1.

```
   99
 - 18
 ----
   81
```

Examples:
99 – 64 =
Subtract 6 from 9: 9 – 6 = 3
Subtract 4 from 9: 9 – 4 = 5

```
   99
 - 64
 ----
   35
```

This even works for larger numbers:

9999 – 2438 =
Subtract 2 from 9: 9 – 2 = 7
Subtract 4 from 9: 9 – 4 = 5
Subtract 3 from 9: 9 – 3 = 6
Subtract 8 from 9: 9 – 8 = 1

```
   9999
 - 2438
 ------
   7561
```

This kind of subtraction can be done from left to right and the answer can be written directly.

There is only <u>one variation</u> that you need to be careful of – what if the number being subtracted has fewer digits than the number with 9s?

9999 – 32 =
For the purpose of calculation, you should think of 32 as 0032 (meaning, put as many leading 'zeroes' so that both the numbers have same number of digits and then follow the same method).
9999 – 0032 =
Subtract 0 from 9: 9 – 0 = 9
Subtract 0 from 9: 9 – 0 = 9
Subtract 3 from 9: 9 – 3 = 6
Subtract 2 from 9: 9 – 2 = 7

```
   9999
 - 0032
 ------
   9967
```

Exercise:

1) 99 − 54 =

```
   9 9
 − 5 4
 ┌─┬─┐
 └─┴─┘
```

2) 99 − 25 =

```
   9 9
 − 2 5
 ┌─┬─┐
 └─┴─┘
```

3) 99 − 3 =

```
   9 9
 − 0 3
 ┌─┬─┐
 └─┴─┘
```

4) 999 − 524 =

```
   9 9 9
 − 5 2 4
 ┌─┬─┬─┐
 └─┴─┴─┘
```

5) 999 − 333 =

```
   9 9 9
 − 3 3 3
 ┌─┬─┬─┐
 └─┴─┴─┘
```

6) 999 − 274 =

```
   9 9 9
 − 2 7 4
 ┌─┬─┬─┐
 └─┴─┴─┘
```

7) 9999 − 1628 =

```
   9 9 9 9
 − 1 6 2 8
 ┌─┬─┬─┬─┐
 └─┴─┴─┴─┘
```

8) 9999 − 518 =

```
   9 9 9 9
 − 0 5 1 8
 ┌─┬─┬─┬─┐
 └─┴─┴─┴─┘
```

9) 99999 − 52376 =

```
   9 9 9 9 9
 − 5 2 3 7 6
 ┌─┬─┬─┬─┬─┐
 └─┴─┴─┴─┴─┘
```

10) 99999 − 235 =

```
   9 9 9 9 9
 − 0 0 2 3 5
 ┌─┬─┬─┬─┬─┐
 └─┴─┴─┴─┴─┘
```

Exercise: In the questions below, write down the answer directly. In some cases the hint is given.

1) 99 − 21   =   (do 9 − 2 and then 9 − 1)

2) 99 − 13   =

3) 99 − 7   =   (9 − 0; 9 − 7)

4) 999 − 123   =

5) 999 − 798   =

6) 999 − 23   =   (9 − 0, 9 − 2; 9 − 3)

7) 999 − 8   =   (9 − 0; 9 − 0; 9 − 8)

8) 9999 − 12   =

9) 9999 − 3418 =

10) 99999 − 5143 =

11) 99999 − 123 =

Shortcut 4: Subtracting a number from a multiple of 10

This is an application of shortcut #3.

> **Rule**
> Example is subtracting some number from 100, 1000 or a higher multiple of 10. The trick is to think of 100 as 1 + 99.
> So when you are subtracting a number from 100, you instead subtract from 99 and then add 1 to the result.
> Subtracting from 99 is very easy as we have seen in shortcut #3. It can be done left to right with no borrowing.

Illustration

Calculate 1000 – 735
= 1 + 999 – 735
= 1 + 264 [do this from left to right (9 – 7; 9 – 3; 9 – 5)]
= 265

Examples:

100 – 24
= 1 + 99 – 24
= 1 + 75
(9 – 2; 9 – 4)
= 76

100 – 59
= 1 + 99 – 59
= 1 + 40
(9 – 5; 9 – 9)
= 41

10000 – 125
= 1 + 9999 – 125
= 1 + 9874
(Think of 125 as 0125 and then you do:
9 – 0; 9 – 1; 9 – 2; 9 – 5)
= 9875

1000 – 628
= 1 + 999 – 628
= 1 + 371 (9 – 6; 9 – 2; 9 – 8)
= 372

100000 – 78933
= 1 + 99999 – 78933
= 1 + 21066
(9 – 7; 9 – 8; 9 – 9; 9 – 3; 9 – 3)
= 21067

100000 – 109
= 1 + 99999 – 109
= 1 + 99890
(9 – 0; 9 – 0; 9 – 1; 9 – 0; 9 – 9)
= 99891

100000 – 2591
= 1 + 99999 – 2591
= 1 + 97408
(9 – 0 ; 9 – 2; 9 – 5; 9 – 9; 9 – 1)
= 97409

100000 – 301
= 1 + 99999 – 301
= 1 + 99698
(9 – 0; 9 – 0; 9 – 0; 9 – 3; 9 – 0; 9 – 1)
=99699

1000000 – 301
= 1 + 999999 – 301
= 999699

Exercise:

1) 100 − 36
= 1 + ☐ − 36
= 1 + ☐
= ☐

2) 100 − 78
= 1 + ☐ − 78
= 1 + ☐
   (9 - 7 ; 9 - 8)
= ☐

3) 1000 − 122
= ☐ + ☐ − 122
= 1 + ☐
= ☐

4) 1000 − 429
= ☐ + ☐ − 429
= ☐ + ☐
= ☐

5) 1000 − 981
= 1 + ☐ − ☐
= ☐ + ☐
= ☐

6) 1000 − 595
= ☐ + 999 − ☐
= ☐ + ☐
= ☐

7) 10000 − 5892
= 1 + ☐ − ☐
= 1 + ☐
= ☐

8) 100000 − 282
= 1 + ☐ − ☐
= 1 + ☐
= ☐

9)   10000 − 2542

= ☐ + ☐ − ☐
= ☐ + ☐
= ☐

---

10)   10000 − 292

= ☐ + ☐ − ☐
= ☐ + ☐
= ☐

Shortcut 5: Split the adder

Rule
When adding two numbers, split one of the numbers to simplify the addition process. This is especially useful if you have one number close to a multiple of 10. Shift enough quantity from one number to make the other number a multiple of 10.

Illustration

Suppose you have to do **197 + 67**
Since 197 is 3 less than 200, we split 67 into 3 + 64
Then 197 + 67
= 197 + 3 + 64
= 200 + 64
= 264

The whole idea is to convert the standard addition of the numbers into an easier form.

Examples:

94 + 39
94 is 6 less than 100
= 94 + 6 + 33
= 100 + 33
= 133

395 + 129
395 is 5 less than 400
= 395 + 5 + 124
= 400 + 124
= 524

1194 + 510
1194 is 6 less than 1200
= 1194 + 6 + 504
= 1200 + 504
= 1704

1094 + 558
1094 is 6 less than 1100
= 1094 + 6 + 552
= 1100 + 552
= 1652

691 + 49
691 is 9 less than 700
= 691 + 9 + 40
= 700 + 40
= 740

789 + 111
789 is 11 less than 800
= 789 + 11 + 100
= 800 + 100
= 900

988 + 219
988 is 12 less than 1000
= 988 + 12 + 207
= 1000 + 207
= 1207

1087 + 276
1087 is 13 less than 1100
= 1087 + 13 + 263
= 1100 + 263
= 1363

Exercise:
In each of the cases shift enough of one number to the other so that the resulting number is an easy multiple of 10.

Solve the following addition problems :

1)   93   +   88   =

How much is needed to make 93 a multiple of 10 = ☐

= 93 + ☐ + ☐

= ☐ + ☐

= ☐

2)   192   +   49   =

How much is needed to make 192 a multiple of 10 = ☐

= 192 + ☐ + ☐

= ☐ + ☐

= ☐

3)   298   +   65   =   298   +   ☐   +   ☐

= ☐ + ☐

= ☐

4)   992   +   79   =   992   +   ☐   +   ☐

= 1000 + ☐

= ☐

5)   797   +   55   =   797   +   ☐   +   ☐

= 800 + ☐

= ☐

23

6) 2095 + 29 = 2095 + 5 + ☐
         = ☐ + ☐
         = ☐

7) 1089 + 221 = 1089 + ☐ + ☐
         = ☐ + ☐
         = ☐

8) 9892 + 288 = ☐ + ☐ + ☐
         = ☐ + ☐
         = ☐

9) 15990 + 4010 = ☐ + ☐ + ☐
         = ☐ + ☐
         = ☐

10) 2998 + 124 = ☐ + ☐ + ☐
         = ☐ + ☐
         = ☐

Shortcut 6: Add in parts

Rule
Add the corresponding place values of the numbers being added separately. This is done by splitting up the numbers into convenient parts for easier addition.

Illustration

Suppose you have to do **123 + 46**
Think of 123 as 120 + 3 and 46 as 40 + 6
Then 123 + 46
= 120 + 3 + 40 + 6
Add 120 and 40 and separately add 3 and 6
= 120 + 40 + 3 + 6
= 160 + 9
= 169

Examples:

58 + 24
58 = 50 + 8
24 = 20 + 4
58 + 24 = 50 + 20 + 8 + 4
= 70 + 12
= 82

38 + 76
38 = 30 + 8
76 = 70 + 6
38 + 76 = 30 + 70 + 8 + 6
= 100 + 14
= 114

92 + 44
92 = 90 + 2
44 = 40 + 4
92 + 44 = 90 + 40 + 2 + 4
= 130 + 6
= 136

264 + 384
= 200 + 60 + 4 + 300 + 80 + 4
= 200 + 300 + 60 + 80 + 4 + 4
= 500 + 140 + 8
= 500 + 148
= 648

548 + 174
= 500 + 40 + 8 + 100 + 70 + 4
= 500 + 100 + 40 + 70 + 8 + 4
= 600 + 110 + 12
= 710 + 12
= 722

1922 + 220
= 1900 + 22 + 200 + 20
= 1900 + 200 + 22 + 20
= 2100 + 42
= 2142

Examples:

723 + 28
723 = 700 + 20 + 3
28 = 20 + 8
723 + 28 = 700 + 40 + 11
= 740 + 11
= 751

905 + 69
= 900 + 5 + 60 + 9
= 900 + 60 + 9 + 5
= 900 + 60 + 14
= 900 + 74
= 974

162 + 34
162 = 100 + 60 + 2
34 = 30 + 4
162 + 34 = 100 + 90 + 6
= 190 + 6
= 196

583 + 117
583 = 500 + 80 + 3
117 = 100 + 10 + 7
= 600 + 90 + 10
= 700

Exercise:

Solve the following addition problems by adding the corresponding place values separately.

1) 45 + 37 =

        45 = 40 + ☐

        37 = 30 + ☐

  45 + 37 = 40 + 30 + ☐ + ☐

          = ☐ + ☐

          = ☐

2) 75 + 18 =

        75 = 70 + ☐

        18 = 10 + ☐

  75 + 18 = ☐ + ☐

          = ☐

3) 38 + 27 =

        38 = ☐ + ☐

        27 = ☐ + ☐

  38 + 27 = ☐ + ☐

          = ☐

4)   198 + 23 =

   198 = 100 + ☐ + 8
   23 =   20 + 3

   198 + 23 = 100 + ☐ + ☐ + ☐ + ☐
            = ☐ + ☐ + ☐
            = ☐ + ☐
            = ☐

5)   625 + 452 =

   625 = 600 + ☐ + ☐
   452 = 400 + ☐ + 2

   625 + 452 = 600 + 400 + ☐ + 50 + ☐ + 2
             = ☐ + 70 + ☐
             = ☐ + ☐
             = ☐

6)   492 + 398 =

   492 = ☐ + ☐ + ☐
   398 = 300 + ☐ + ☐

   492 + 398 = 400 + ☐ + 90 + ☐ + 2 + ☐
             = ☐ + ☐ + 10
             = ☐ + ☐
             = ☐

7) 1125 + 527 =

1125 = ☐ + ☐ + ☐ + ☐

527 = ☐ + ☐ + ☐

1125 + 527 = ☐ + ☐ + ☐ + ☐

= ☐ + ☐

= ☐

8) 1825 + 954 =

1825 = ☐ + ☐ + ☐ + ☐

954 = ☐ + ☐ + ☐

1825 + 954 = ☐ + ☐ + ☐ + ☐

= ☐ + ☐

= ☐

9) 9892 + 1522 =

9892 = ☐ + ☐ + ☐ + ☐

1522 = ☐ + ☐ + ☐ + ☐

9892 + 1522 = ☐ + ☐ + ☐ + ☐

= ☐ + ☐

= ☐

10) 19895 + 5820 =

19895 = 19000 + ☐ + ☐ + ☐

5820 = ☐ + ☐ + ☐

19895 + 5820 = ☐ + ☐ + ☐ + ☐

= ☐ + ☐

= ☐

Shortcut 7: Add more or less than needed

Rule
Instead of adding the numbers directly, add a slightly different number that is close to one of the numbers and then adjust the answer.

Illustration

Suppose you have to do **459 + 97**.
Instead of adding 97, add 100. Because 100 is 3 more than 97. So from the result we subtract 3.
459 + 97
= 459 + 100 − 3
= 559 − 3
= 556

Examples:

18 + 9
= 18 + 10 − 1
= 28 − 1
= 27

19 + 18
= 19 + 20 − 2
= 39 − 2
= 37

468 + 104
= 468 + 100 + 4
= 568 + 4
= 572

75 + 52
= 75 + 50 + 2
= 125 + 2
= 127

98 + 47
= 100 + 47 − 2
= 147 − 2
= 145

236 + 49
= 236 + 50 − 1
= 286 − 1
= 285

547 + 39
= 547 + 40 − 1
= 587 − 1
= 586

85 + 38
= 85 + 40 − 2
= 125 − 2
= 123

539 + 388
= 539 + 400 − 12
= 939 − 12
= 927

239 + 294
= 239 + 300 − 6
= 539 − 6
= 533

Exercise:

1) 428 + 95
= ☐ + 100 − 5
= ☐ − 5
= ☐

2) 874 + 88
= ☐ + 100 − 12
= ☐ − 12
= ☐

3) 652 + 99
= ☐ + ☐ − 1
= ☐ − 1
= ☐

4) 219 + 49
= 219 + ☐ − ☐
= ☐ − ☐
= ☐

5) 640 + 39
= ☐ + ☐ − ☐
= ☐ − ☐
= ☐

6) 550 + 204
= ☐ + ☐ + ☐
= ☐ + ☐
= ☐

7) 890 + 59
= ☐ + ☐ − ☐
= ☐ − ☐
= ☐

8) 989 + 108
= ☐ + ☐ + ☐
= ☐ + ☐
= ☐

9) 784 + 92

= ☐ + ☐ − ☐

= ☐ − ☐

= ☐

10) 4462 + 998

= ☐ + ☐ − ☐

= ☐ − ☐

= ☐

Shortcut 8: Subtract more than necessary

Rule
Instead of subtracting a number as is, subtract a different but more convenient number and then adjust the answer.

Illustration

Suppose you have to do 176 − 98
In order to subtract 98, it is better to subtract 100 and then add 2.
This works because 98 = 100 − 2.
Then 176 − 98
= 176 − 100 + 2
= 76 + 2
= 78

The main idea is to subtract an easy number that is also close to the original number being subtracted.

Examples:

82 − 9
= 82 − 10 + 1
= 72 + 1
= 73

75 − 28
= 75 − 30 + 2
= 45 + 2
= 47

49 − 18
= 49 − 20 + 2
= 29 + 2
= 31

129 − 47
= 129 − 50 + 3
= 79 + 3
= 82

2120 − 128
= 2120 − 130 + 2
= 1990 + 2
= 1992

66 − 17
= 66 − 20 + 3
= 46 + 3
= 49

42 − 27
= 42 − 30 + 3
= 12 + 3
= 15

362 − 185
= 362 − 200 + 15
= 162 + 15
= 177

951 − 785
= 951 − 800 + 15
= 151 + 15
= 166

7629 − 174
= 7629 − 200 + 26
= 7429 + 26
= 7455

Exercise:

1) 93 − 8
= ☐ − 10 + 2
= ☐ + 2
= ☐

2) 89 − 38
= 89 − ☐ + 2
= ☐ + 2
= ☐

3) 134 − 36
= 134 − ☐ + 4
= ☐ + 4
= ☐

4) 118 − 27
= 118 − ☐ + 3
= ☐ + 3
= ☐

5) 190 − 46
= 190 − ☐ + ☐
= ☐ + ☐
= ☐

6) 482 − 36

= 482 − ☐ + ☐

= ☐ + ☐

= ☐

7) 362 − 48

= ☐ − ☐ + ☐

= ☐ + ☐

= ☐

8) 281 − 29

= ☐ − ☐ + ☐

= ☐ + ☐

= ☐

9) 1200 − 987

= ☐ − ☐ + ☐

= ☐ + ☐

= ☐

10) 14000 − 7995

= ☐ − ☐ + ☐

= ☐ + ☐

= ☐

Shortcut 9: Adding a series of consecutive numbers from 1 to 'n'

> **Rule**
> To add a series of consecutive numbers from 1 to 'n', just do $\frac{n \times (n+1)}{2}$.

Illustration

What is 1+2+3+4+.....+19+20 (sum of first 20 numbers)
n = 20 , n + 1 = 21
$$\frac{n \times (n+1)}{2}$$
$= \frac{(20 \times 21)}{2}$
$= 10 \times 21$
$= 210$

Examples:

What is 1+2+3+... +9+10
n = 10, n + 1 = 11
$$\frac{n \times (n+1)}{2}$$
$= \frac{(10 \times 11)}{2}$
$= 5 \times 11$
$= 55$

What is 1+2+3+... +24+25
n = 25, n + 1 = 26
$$\frac{n \times (n+1)}{2}$$
$= \frac{(25 \times 26)}{2}$
$= 25 \times 13$
$= 325$

What is 1+2+3+... +11+12
n = 12, n + 1 = 13
$$\frac{n \times (n+1)}{2}$$
$= \frac{(12 \times 13)}{2}$
$= 6 \times 13$
$= 78$

What is 1+2+3+... +14+15
$= \frac{(15 \times 16)}{2}$
$= 15 \times 8$
$= 120$

What is 1+2+3+... +17+18
n = 18, n + 1 = 19
$$\frac{n \times (n+1)}{2}$$
$= \frac{(18 \times 19)}{2}$
$= 9 \times 19$
$= 171$

What is 1+2+3+... +44+45
$= \frac{(45 \times 46)}{2}$
$= 45 \times 23$
$= 1035$

What is 1+2+3+... +29+30
$= \frac{(30 \times 31)}{2}$
$= 15 \times 31$
$= 465$

Exercise:

1) $1+2+3+4+\ldots+39+40$

$= \dfrac{(40 \times 41)}{2}$

$= \square \times \square$

$= \square$

2) $1+2+3+4+\ldots+49+50$

$= \dfrac{(\square \times 51)}{2}$

$= \square \times \square$

$= \square$

3) $1+2+3+4+\ldots+54+55$

$= \dfrac{(55 \times \square)}{2}$

$= \square \times \square$

$= \square$

4) $1+2+3+4+\ldots+59+60$

$= \dfrac{(\square \times \square)}{2}$

$= \square \times \square$

$= \square$

5) $1+2+3+4+\ldots+69+70$

$= \dfrac{(\square \times \square)}{2}$

$= \square \times \square$

$= \square$

6) $1+2+3+4+\ldots+64+65$

$= \dfrac{(\square \times \square)}{\square}$

$= \square \times \square$

$= \square$

7) $1+2+3+4+\ldots+27+28$

$= \dfrac{(\square \times \square)}{\square}$

$= \square \times \square$

$= \square$

8) $1+2+3+4+\ldots+47+48$

$= \dfrac{(\square \times \square)}{2}$

$= \square \times \square$

$= \square$

9) 1+2+3+4+......+99+100

= (☐ X ☐) / ☐

= ☐ X ☐

= ☐

10) 1+2+3+4+....+199+200

= (☐ X ☐) / ☐

= ☐ X ☐

= ☐

Shortcut 10: Sum of first 'n' even numbers

Rule
To calculate the sum of first 'n' even numbers, find the value of:

n X (n + 1)

Illustration

2 + 4 + 6 + 8 + 10 =

Here n = 5 because we want to find the sum of first 5 even numbers.
n = 5
n + 1 = 6

2 + 4 + 6 + 8 + 10 =

= 5 X 6

= 30

Examples:
2 + 4 + 6 + ... + 20 =

n = 10
n + 1 = 11

2 + 4 + 6 + ... + 20 =

= 10 X 11

= 110

Sidebar

How do you find the value of 'n'?

Example: if you have to find
2 + 4 + 6 + ... 54 + 56 , then how do you find how many numbers are there? What is the value of 'n'?

Look at the largest even number in the series and divide that number by 2. That is the value of 'n'.

Exercise:

1) 2 + 4 + 6 + ….. + 16 =

   n = 8
   n + 1 = [ ]

   2 + 4 + ….. + 16 =
   = [ ] x [ ]
   = [ ]

---

2) 2 + 4 + 6 + ….. + 26 =

   n = 13
   n + 1 = 14

   2 + 4 + 6 + ….. + 26 =
   = [ ] x [ ]
   = [ ]

---

3) 2 + 4 + ….. + 38 + 40 =

   n = [ ]
   n + 1 = [ ]

   2 + 4 + ….. + 38 + 40 =
   = [ ] x [ ]
   = [ ]

4) 2 + 4 + ... + 48 + 50 =

n = ☐
n + 1 = ☐

2 + 4 + ... + 48 + 50 =
= ☐ × ☐
= ☐

5) 2 + 4 + ... + 58 + 60 =

n = ☐
n + 1 = ☐

2 + 4 + ... + 58 + 60 =
= ☐ × ☐
= ☐

6) 2 + 4 + ... + 98 + 100 =
= ☐ × ☐
= ☐

7) 2 + 4 + ... + 148 + 150 =

= ☐ X ☐

= ☐

8) 2 + 4 + ... + 198 + 200 =

= ☐ X ☐

= ☐

9) 2 + 4 + ... + 498 + 500 =

= ☐ X ☐

= ☐

10) 2 + 4 + ... + 998 + 1000 =

= ☐ X ☐

= ☐

Shortcut 11: Sum of first 'n' odd numbers

Rule
To calculate the sum of first 'n' odd numbers, find the value of:

n^2

Illustration

1 + 3 + 5 + 7 + 9 =

Here n = 5 because we want to find the sum of first 5 odd numbers.
n = 5

n^2 = 25
1 + 3 + 5 + 7 + 9 = 25

Examples:
1 + 3 + ... + 13 + 15 =
n = 8

n^2 = 64
1 + 3 + ... + 13 + 15 = 64

1 + 3 + ... + 23 + 25 =
n = 13

n^2 = 169
1 + 3 + ... + 23 + 25 = 169

Sidebar

How do you find the value of 'n' ?

Example: if you have to find
1 + 3 + ... + 97 + 99, then how do you find how many numbers are there? What is the value of 'n'?

Look at the largest odd number, add 1 to it and divide the result by 2. That is the value of 'n'.
In the case above, n = (99 + 1) ÷ 2
= 100 ÷ 2
= 50

Exercise:

1) $1 + 3 + \ldots + 11 + 13 =$

     $n = 7$
     $n^2 =$ ☐

     $1 + 3 + \ldots + 11 + 13 =$
     $=$ ☐

---

2) $1 + 3 + \ldots + 19 + 21 =$

     $n = 11$
     $n^2 =$ ☐

     $1 + 3 + \ldots + 19 + 21 =$
     $=$ ☐

---

3) $1 + 3 + \ldots + 33 + 35 =$

     $n =$ ☐
     $n^2 =$ ☐

     $1 + 3 + \ldots + 33 + 35 =$
     $=$ ☐

4) $1 + 3 + \ldots + 27 + 29 =$

   n = 15

   n² = ☐

   $1 + 3 + \ldots + 27 + 29 =$

   = ☐

---

5) $1 + 3 + \ldots + 43 + 45 =$

   n = ☐

   n² = ☐

   $1 + 3 + \ldots + 43 + 45 =$

   = ☐

---

6) $1 + 3 + \ldots + 77 + 79 =$

   n = ☐

   n² = ☐

   $1 + 3 + \ldots + 77 + 79 =$

   = ☐

7) $1 + 3 + \ldots + 97 + 99 =$

   $n = \boxed{}$
   $n^2 = \boxed{}$

   $1 + 3 + \ldots + 97 + 99 =$
   $= \boxed{}$

---

8) $1 + 3 + \ldots + 147 + 149 =$

   $n = \boxed{}$
   $n^2 = \boxed{}$

   $1 + 3 + \ldots + 147 + 149 =$
   $= \boxed{}$

---

9) $1 + 3 + \ldots + 197 + 199 =$

   $n = \boxed{}$
   $n^2 = \boxed{}$

   $2 + 4 + \ldots + 197 + 199 =$
   $= \boxed{}$

---

10) $1 + 3 + \ldots + 497 + 499 =$

    $n = \boxed{}$
    $n^2 = \boxed{}$

    $2 + 4 + \ldots + 497 + 499 =$
    $= \boxed{}$

Shortcut 12: Multiplying a number by a multiple of 10

Rule
For example, multiplying a number by 10, 100, 10000 etc.
Count the number of 'zeroes' in the multiple of 10 by which you are multiplying and put that many 'zeroes' at the end of the number being multiplied.

Illustration

What is 23 X 100
100 has two zeroes. Put two zeroes at the end of 23.
23 X 100
= 2300

Examples:

27 X 10
10 has one zero
27 X 10
= 270

43 x 100
100 has two zeroes
43 X 100
= 4300

77 X 1000
1000 has three zeroes
77 X 1000
= 77000

320 X 100
100 has two zeroes
320 X 100
= 32000
(320 already has a 'zero' at the end so we put two more 'zeroes' after that.)

700 X 100
= 70000 (put two 'zeroes' after 700)

324 X 1000
= 324000

708 X 1000
= 708000

1000 X 1000
= 1000000 (three 'zeroes' are already there and we put three more 'zeroes' after them.)

Exercise:

1) 18 X 10
   = ☐

2) 24 X 10
   = ☐

3) 60 X 10
   = ☐

4) 84 X 100
   = ☐

5) 100 X 100
   = ☐

6) 102 X 10
   = ☐

7) 1010 X 100
   = ☐

8) 38 X 1000
   = ☐

9) 1245 X 100
   = ☐

10) 6000 X 1000
    = ☐

11) 84000 X 1000
    = ☐

12) 1000 X 1000
    = ☐

13) 70000 X 10000
    = ☐

14) 1000000000 X 100000
    = ☐

Shortcut 13: Multiplying decimal numbers by a multiple of 10 (e.g. 10,100,1000 etc)

Rule
When any number with decimal is multiplied with 10 the decimal moves one place to the right.
If any number with decimal is multiplied with 100 or 1000 the decimal moves two or three places to the right.
The rule is to shift the decimal to the right by as many places as there are 'zeroes' in the multiple of 10 that is being multiplied.

Illustration

Multiply 6.253 X 10
Now move decimal point one place to the right
= 62.53

Examples:

1.25 X 10
Move decimal point one place to the right
= 12.5

0.69 X 10
Move decimal point one place to the right
= 6.9

45.16 X 10
Move decimal point one place to the right
=451.6

391.592 X 10
Move decimal point one place to the right
= 3915.92

21.753 X 100
Move decimal point two places to the right
= 2175.3

8.28 X 1000
Move decimal point three places to the right
= 8280
Because 8.28 has two places after decimal and we need to shift by three places, we have to add a 'zero'.

394.62 X 1000
Move decimal point three places to the right
= 394620

52.446 X 100
Move decimal point two places to the right
= 5244.6

4.25 X 100
Move decimal point two places to the right
= 425

28.2589 X 1000
Move decimal point three places to the right
= 28258.9

85.374 X 10000
Move decimal point four places to the right
= 853740

735.56 X 1000
Move decimal point three places to the right
= 735560

Exercise:

1)   0.69   X   10

Move decimal point one place to the right.

= [   ]

---

2)   28.359   X   10

Move decimal point one place to the right.

= [   ]

---

3)   107.42   X   10

= [   ]

---

4)   580.31   X   10

= [   ]

---

5)   62.23   X   100

= [   ]

---

6)   485.202   X   100

= [   ]

7) 43.8 X 100
= [ ]

8) 82.151 X 100
= [ ]

9) 92.4976 X 1000
= [ ]

10) 72.8574 X 1000
= [ ]

11) 217.753 X 1000
= [ ]

12) 808.78 X 1000
= [ ]

Shortcut 14: Dividing by 10 or higher multiples of 10

> **Rule**
> Dividing by 10 is same as putting a decimal in the number after leaving one digit to the right. If you are dividing by 100, leave two digits to the right. So the simple rule is:
>
> Count how many of 'zeroes' are there in the number you are dividing by and put a decimal after that many digits to the right in the number being divided.
>
> If the number being divided already has decimal, then shift the decimal to the left by as many spaces as there are 'zeroes' in the multiple of 10.
>
> There is a slight <u>variation</u> to this if the number being divided ends in 'zeroes'. We will cover it in the examples below:

Illustration

Calculate $\frac{73}{10}$

10 has one zero, so we will put a decimal with one digit to the right of 73.

$\frac{73}{10} = 7.3$

Examples:

Calculate $\frac{73}{100}$

100 has two zeroes. So we will put a decimal with two digits to the right.

$\frac{73}{100} = 0.73$

Note: Since 73 has two digits and we have to put a decimal with two digits to the right, it means we are putting a decimal before 73. So the answer is .73.
The right way to say it is 0.73, because <u>we should always have a digit to the left of the decimal.</u>

Calculate $\frac{730}{100}$

We will put a decimal with two digits to the right.

So $\frac{730}{100}$ = 7.30 which is same as 7.3

Calculate $\frac{73}{1000}$

Now 1000 has three zeroes. So we have to put a decimal with three digits to the right of the number. Since 73 has only two digits, we will put a 'zero' before it and then put a decimal so that there are three digits to the right of the decimal.

$\frac{73}{1000}$ = 0.073 (<u>we should always have a digit to the left of the decimal.</u>)

Examples:

What if there are zeroes to the right?

Calculate $\frac{730}{10}$

The rule still holds. Put a decimal with one digit to the right. That means:

$$\frac{730}{10} = 73.0$$

Since a zero after decimal with no other digit to its right does not carry any value, 73.0 is same as 73

Calculate $\frac{98}{10}$

10 has one zero, so we will put a decimal with one digit to the right of 98

$$\frac{98}{10} = 9.8$$

Calculate $\frac{98}{100}$

100 has two zeroes. So we will put a decimal with two digits to the right.

$$\frac{98}{100} = 0.98$$

Calculate $\frac{98}{1000}$

1000 has three zeroes. So we have to put a decimal with three digits to the right of the number. Since 98 has only two digits, we will put a 'zero' before it and then put a decimal so that there are three digits to the right of the decimal.

$$\frac{98}{1000} = 0.098$$

Calculate $\frac{980}{10}$

Put a decimal with one digit to the right. That means:

$$\frac{980}{10} = 98.0 \text{ which is same as } 98$$

Calculate $\frac{980}{100}$

We will put a decimal with two digits to the right.

So $\frac{980}{100} = 9.80$ which is same as 9.8

Examples:

Calculate $\frac{980}{1000}$

Now 1000 has three 'zeroes'. So we have to put a decimal with three digits to the right of the number. Since 980 has three digits, we will put a 'zero' before it.

$\frac{980}{1000}$ = .980 which is same as 0.98 (note that we have removed the <u>trailing 'zero'</u>).

Calculate $\frac{9.8}{10}$

9.8 has a decimal already. 10 has <u>one</u> 'zero'. So we will shift the decimal one space to the left.

$\frac{9.8}{10}$ = 0.98

Calculate $\frac{9.8}{1000}$

1000 has three 'zeroes', so we will shift the decimal three places to the left.

$\frac{9.8}{1000}$ = 0.0098

Exercise:

1) $\dfrac{23}{10}$ = ☐

2) $\dfrac{49}{10}$ = ☐

3) $\dfrac{123}{10}$ = ☐

4) $\dfrac{109}{10}$ = ☐

5) $\dfrac{180}{10}$ = ☐

6) $\dfrac{23}{100}$ = ☐

7) $\dfrac{49}{100}$ = ☐

8) $\dfrac{123}{100}$ = ☐

9) $\dfrac{109}{100}$ = ☐

10) $\dfrac{180}{100}$ = ☐

11) $\dfrac{543}{10}$ = ☐

12) $\dfrac{4929}{10}$ = ☐

13) $\dfrac{543}{100}$ = ☐

14) $\dfrac{4929}{100}$ = ☐

15) $\dfrac{23}{1000}$ = ☐

16) $\dfrac{49}{1000}$ = ☐

17) $\dfrac{123}{1000}$

= ☐

18) $\dfrac{109}{1000}$

= ☐

19) $\dfrac{543}{1000}$

= ☐

20) $\dfrac{4929}{1000}$

= ☐

21) $\dfrac{17.25}{10}$

= ☐

22) $\dfrac{239.8}{1000}$

= ☐

23) $\dfrac{9999.9}{100}$

= ☐

24) $\dfrac{0.241}{100}$

= ☐

25) $\dfrac{999.9}{10}$

= ☐

Shortcut 15: Multiplying numbers between 10 and 19

Rule
1. Add the <u>unit's</u> place digit of either of the numbers to the other number.
2. Multiply this result by 10.
3. Multiply the unit's place digits of both the numbers.
4. Add the answers you get in step 2 and step 3.

Illustration

Calculate 16 X 18
1. 16 + 8 = 24 (add the unit's place digit of 18 to the other number, which is 16.)
2. 24 x 10 = 240 (multiply the result of step 1 by 10.)
3. 6 x 8 = 48 (multiply the unit's place digits of both the numbers.)
4. 240 + 48 = 288 (add the answers you get in step 2 and 3.)

So: 16 x 18 = 288

Examples:

11 X 12
1. 11 + 2 = 13
2. 13 x 10 = 130
3. 1 x 2 = 2
4. 130 + 2 = 132

11 x 12 = 132

13 X 17
1. 13 + 7 = 20
2. 20 x 10 = 200
3. 3 x 7 = 21
4. 200 + 21 = 221

13 x 17 = 221

12 X 15
1. 12 + 5 = 17
2. 17 x 10 = 170
3. 2 x 5 = 10
4. 170 + 10 = 180

12 x 15 = 180

Exercise:

1) 16 X 17

    16 + 7 = ☐

    ☐ X 10 = ☐

    6 X 7 = ☐

    ☐ + ☐ = ☐

---

2) 18 X 14

    18 + 4 = ☐

    ☐ X 10 = ☐

    8 X 4 = ☐

    ☐ + ☐ = ☐

---

3) 13 X 11

    13 + 1 = ☐

    ☐ X 10 = ☐

    3 X 1 = ☐

    ☐ + ☐ = ☐

---

4) 16 X 14

    16 + 4 = ☐

    ☐ X 10 = ☐

    6 X 4 = ☐

    ☐ + ☐ = ☐

5)   18   X   12

       18   +   2   =   [   ]

   [   ]   X   10   =   [   ]

       8   X   2   =   [   ]

   [   ]   +   [   ]   =   [   ]

---

6)   18   X   17

   [   ]   +   [   ]   =   [   ]

   [   ]   X   10   =   [   ]

   [   ]   X   [   ]   =   [   ]

   [   ]   +   [   ]   =   [   ]

---

7)   19   X   12

   [   ]   +   [   ]   =   [   ]

   [   ]   X   10   =   [   ]

   [   ]   X   [   ]   =   [   ]

   [   ]   +   [   ]   =   [   ]

---

8)   15   X   13

   [   ]   +   [   ]   =   [   ]

   [   ]   X   [   ]   =   [   ]

   [   ]   X   [   ]   =   [   ]

   [   ]   +   [   ]   =   [   ]

9)   13   X   12

□ + □ = □
□ X □ = □
□ X □ = □
□ + □ = □

10)   19   X   11

□ + □ = □
□ X □ = □
□ X □ = □
□ + □ = □

Shortcut 16: Subtract in Steps

Rule
Split the number being subtracted into more than one part to simplify the subtraction.

Illustration

Calculate 987 – 32
Instead of subtracting 32, first subtract 30 and then subtract 2
987 – 32
= 987 – 30 – 2
= 957 – 2
= 955

Examples:

65 – 12
= 65 – 10 – 2
= 55 – 2
= 53

170 – 31
= 170 – 30 – 1
= 140 – 1
= 139

863 – 128
= 863 – 100 – 28
= 763 – 28
= 763 – 20 – 8 (spilt 28 into 20 and 8)
= 743 – 8
= 735

243 – 28
= 243 – 20 – 8
= 223 – 8
= 215

36 – 12
= 36 – 10 – 2
= 26 – 2
= 24

367 – 229
= 367 – 200 – 29
= 167 – 20 – 9
= 147 – 9
= 138

98 – 16
= 98 – 10 – 6
= 88 – 6
= 82

863 – 426
= 863 – 400 – 20 – 6
= 463 – 20 – 6
= 443 – 6
= 437

278 – 22
= 278 – 20 – 2
= 258 – 2
= 256

656 – 338
= 656 – 300 – 30 – 8
= 356 – 30 – 8
= 326 – 8
= 318

Exercise:

1) 334 − 26
= 334 − ☐ − 6
= ☐ − 6
= ☐

2) 861 − 43
= 861 − ☐ − 3
= ☐ − 3
= ☐

3) 594 − 76
= ☐ − ☐ − ☐
= ☐ − ☐
= ☐

4) 677 − 54
= ☐ − ☐ − ☐
= ☐ − ☐
= ☐

5) 983 − 37
= 983 − ☐ − ☐
= ☐ − ☐
= ☐

6) 756 − 42
= 756 − ☐ − ☐
= ☐ − ☐
= ☐

7) 656 − 338
= 656 − 300 − 38
= ☐ − ☐ − ☐ (further spilt 38)
= ☐ − ☐
= ☐

8) 953 − 717
= 953 − 700 − ☐
= ☐ − ☐ − ☐
= ☐ − ☐
= ☐

9) 245 − 32
= ☐ − ☐ − ☐
= ☐ − ☐
= ☐

10) 756 − 459
= ☐ − ☐ − ☐ − 9
= ☐ − ☐ − ☐
= ☐ − ☐
= ☐

Shortcut 17: Subtract by simplifying

Rule
Keep subtracting parts of numbers to simplify the calculation.

Illustration

Calculate 1103 – 38
Subtract 3 from both the numbers
1103 – 38
= 1100 – 35
Subtract 5 from both numbers
= 1095 – 30
= 1065

Examples:

65 – 17
Subtract 5 from both numbers
= 60 – 12
Subtract 2 from both numbers
= 58 – 10
= 48

763 – 132
Subtract 3 from both numbers
= 760 – 129
Subtract 20 from both numbers
= 740 – 109
Subtract 9 from both numbers
= 731 – 100
= 631

89 – 17
Subtract 7 from both numbers
= 82 – 10
= 72

189 – 73
Subtract 3 from both numbers
= 186 – 70
= 116

992 – 134
Subtract 30 from both numbers
= 962 – 104
Subtract 4 from both numbers
= 958 – 100
= 858

848 – 242
Subtract 40 from both numbers
= 808 – 202
Subtract 2 from both numbers
= 806 – 200
= 606

287 – 148
Subtract 40 from both numbers
= 247 – 108
Subtract 8 from both numbers
= 239 – 100
= 139

93 – 17
Subtract 7 from both numbers
= 86 – 10
= 76

175 – 48
Subtract 8 from both numbers
= 167 – 40
= 127

598 – 189
Subtract 9 from both numbers
= 589 – 180
Subtract 80 from both numbers
= 509 – 100
= 409

Exercise:

1) 59 − 13
Subtract 3 from both numbers
= ☐ − 10
= ☐

2) 78 − 29
Subtract 9 from both numbers
= ☐ − ☐
= ☐

3) 84 − 37
Subtract ☐ from both numbers
= ☐ − ☐
= ☐

4) 123 − 76
Subtract ☐ from both numbers
= ☐ − ☐
= ☐

5) 149 − 54
Subtract ☐ from both numbers
= 145 − ☐
= ☐

6) 188 − 129
Subtract 9 from both numbers
= ☐ − ☐
Subtract 20 from both numbers
= ☐ − 100
= ☐

7) 473 − 29
Subtract 9 from both numbers
= ☐ − ☐
= ☐

8) 1108 − 149
Subtract ☐ from both numbers
= ☐ − ☐
Subtract ☐ from both numbers
= ☐ − ☐
= ☐

9)   729   –   67
       Subtract [ ] from both numbers
=   [   ] – [   ]
=   [   ]

---

10)   2210 –   164
       Subtract [ ] from both numbers
=   [   ] – [   ]
       Subtract [ ] from both numbers
=   [   ] – [   ]
=   [   ]

Shortcut 18: Multiplying a number by 5

Rule
Multiply the given number by 10 and then divide the result by 2. This is the same as putting a '0' at the end of the number and then divide the number by 2.
This works because 5 = 10 ÷ 2.

Illustration

Calculate 242 X 5
This the same as:
2420 divided by 2
= 1210

Examples:

Calculate 18 X 5
= $\frac{180}{2}$
= 90

Calculate 29 X 5
= $\frac{290}{2}$
= 145

Calculate 31 X 5
= $\frac{310}{2}$
= 155

Calculate 81 X 5
= $\frac{810}{2}$
= 405

Calculate 245 X 5
= $\frac{2450}{2}$
= 1225

Calculate 398 X 5
= $\frac{3980}{2}$
= 1990

Calculate 789 X 5
= $\frac{7890}{2}$
= 3945

Calculate 1853 X 5
= $\frac{18530}{2}$
= 9265

Calculate 8981 X 5
= $\frac{89810}{2}$
= 44905

Calculate 43216 X 5
= $\frac{432160}{2}$
= 216080

Calculate 98989 X 5
= $\frac{989890}{2}$
= 494945

Exercise:

1)   89   X   5

= $\dfrac{89 \times 10}{2}$

= $\dfrac{890}{2}$

= ☐

2)   112   X   5

= $\dfrac{112 \times 10}{2}$

= $\dfrac{\boxed{}}{2}$

= ☐

3)   199   X   5

= $\dfrac{199 \times 10}{2}$

= $\dfrac{\boxed{}}{2}$

= ☐

4)   2189   X   5

= $\dfrac{2189 \times 10}{2}$

= $\dfrac{\boxed{}}{2}$

= ☐

5)   3899   X   5

= $\dfrac{\boxed{}}{2}$

= ☐

6)   5695   X   5

= $\dfrac{\boxed{}}{2}$

= ☐

7)   99854   X   5

= $\dfrac{\boxed{}}{\boxed{}}$

= ☐

8)   19948   X   5

= $\dfrac{\boxed{}}{\boxed{}}$

= ☐

9)   76480   X   5

=   ☐
    ☐

=   ☐

---

10)   29872   X   5

=   ☐
    ☐

=   ☐

Shortcut 19: Dividing a number by 5

Rule
Multiply the number by 2.
Divide the number by 10 (refer shortcut #14).
This is because $5 = \frac{10}{2}$. So dividing by 5 is same as doubling the number and then dividing by 10.

Illustration

Calculate $\frac{43}{5}$

$43 \times 2 = 86$

$\frac{86}{10} = 8.6$

$\frac{43}{5} = 8.6$

Examples:

Calculate $\frac{542}{5}$

$542 \times 2 = 1084$

$\frac{1084}{10} = 108.4$

$\frac{542}{5} = 108.4$

Calculate $\frac{68997}{5}$

$68997 \times 2 = 137994$

$\frac{137994}{10} = 13799.4$

$\frac{68997}{5} = 13799.4$

Calculate $\frac{624}{5}$

$624 \times 2 = 1248$

$\frac{1248}{10} = 124.8$

$\frac{624}{5} = 124.8$

Calculate $\frac{2899}{5}$

$2899 \times 2 = 5798$

$\frac{5798}{10} = 579.8$

$\frac{2899}{5} = 579.8$

Calculate $\frac{76395}{5}$

$76395 \times 2 = 152790$

$\frac{152790}{10} = 15279$

$\frac{76395}{5} = 15279$

Calculate $\frac{59}{5}$

$59 \times 2 = 118$

$\frac{59}{10} = 11.8$

$\frac{118}{5} = 11.8$

Examples:

Calculate $\frac{83}{5}$

83 X 2 = 166

$\frac{83}{10}$ = 16.6

$\frac{166}{5}$ = 16.6

Calculate $\frac{9521}{5}$

9521 X 2 = 19042

$\frac{19042}{10}$ = 1904.2

$\frac{9521}{5}$ = 1904.2

Calculate $\frac{128}{5}$

128 X 2 = 256

$\frac{256}{10}$ = 25.6

$\frac{128}{5}$ = 25.6

Calculate $\frac{349}{5}$

349 X 2 = 698

$\frac{698}{10}$ = 69.8

$\frac{349}{5}$ = 69.8

Calculate $\frac{6342}{5}$

6342 X 2 = 12684

$\frac{12684}{10}$ = 1268.4

$\frac{6342}{5}$ = 1268.4

Calculate $\frac{8112}{5}$

8112 X 2 = 16224

$\frac{16224}{10}$ = 1622.4

$\frac{8112}{5}$ = 1622.4

Exercise:

1) $\dfrac{74}{5}$
= ☐ × 2 = ☐
= ☐ ÷ 10
= ☐

2) $\dfrac{94}{5}$
= ☐ × 2 = ☐
= ☐ ÷ 10
= ☐

3) $\dfrac{112}{5}$
= ☐ × ☐ = ☐
= ☐ ÷ 10
= ☐

4) $\dfrac{459}{5}$
= ☐ × ☐ = ☐
= ☐ ÷ 10
= ☐

5) $\dfrac{891}{5}$
= ☐ × ☐ = ☐
= ☐ ÷ ☐
= ☐

6) $\dfrac{1024}{5}$
= ☐ × 2 = ☐
= ☐ ÷ ☐
= ☐

7) $\dfrac{2921}{5}$
= ☐ × ☐ = ☐
= ☐ ÷ ☐
= ☐

8) $\dfrac{3459}{5}$
= ☐ × ☐ = ☐
= ☐ ÷ ☐
= ☐

9) $\dfrac{86403}{5}$
= ☐ × ☐ = ☐
= ☐ ÷ ☐
= ☐

10) $\dfrac{98492}{5}$
= ☐ × ☐ = ☐
= ☐ ÷ ☐
= ☐

Shortcut 20: Subtract in groups

Rule
Subtracting groups of place values seperately and then adjusting the answers. The key idea is to prevent the 'borrowing' of digits by always subtracting a smaller number from a bigger one.

Illustration

Calculate 483 − 267
Normal subtraction would require us to borrow 1 from 8 to make 13 so that we can subtract 7 from it.
A smarter way is to think of this as:
= 400 − 200 + 83 − 67
= 200 + 16
= 216

Examples:

236 − 62
Think of it as:
= 230 − 60 + 6 − 2
= 170 + 4
= 174

14523 − 7219
= 14000 − 7000 + 500 − 200 + 23 − 19
= 7000 + 300 + 4
= 7304

562 − 328
= 500 − 300 + 62 − 28
= 200 + 34
= 234

1211 − 1009
= 1200 − 1000 + 11 − 9
= 200 + 2
= 202

198 − 63
= 100 + 90 − 60 + 8 − 3
= 100 + 30 + 5
= 135

394 − 122
= 300 − 100 + 90 − 20 + 4 − 2
= 200 + 70 + 2
= 272

784 − 132
= 700 − 100 + 80 − 30 + 4 − 2
= 600 + 50 + 2
= 652

15455 − 8250
= 15000 − 8000 + 400 − 200 + 55 − 50
= 7000 + 200 + 5
= 7205

Exercise:

1)   429   −   123

=   400   −   100   +   ☐   −   ☐

=   ☐   +   ☐

=   ☐

---

2)   146   −   31

=   100   +   ☐   −   ☐

=   100   +   ☐

=   ☐

---

3)   398   −   136

=   ☐ − ☐   +   ☐ − ☐   +   ☐ − ☐

=   ☐ + ☐ + ☐

=   ☐ + ☐

=   ☐

---

4)   567   −   136

=   ☐ − ☐   +   ☐ − ☐   +   ☐ − ☐

=   ☐ + ☐ + ☐

=   ☐ + ☐

=   ☐

5) 755 − 24

= 700 + ☐ − ☐ + ☐ − ☐
= 700 + ☐ + ☐
= ☐ + ☐
= ☐

6) 988 − 126

= 900 − ☐ + ☐ − ☐ + ☐ − ☐
= ☐ + ☐ + ☐
= ☐ + ☐
= ☐

7) 896 − 153

= ☐ − ☐ + ☐ − ☐ + ☐ − ☐
= ☐ + ☐ + ☐
= ☐ + ☐
= ☐

8) 1129 − 112

= 1000 + ☐ − ☐ + ☐ − ☐
= 1000 + ☐ + ☐
= ☐ + ☐
= ☐

9) 1492 − 371

= ☐ + ☐ − ☐ + ☐ − ☐ + ☐ − ☐
= ☐ + ☐ + ☐ + ☐
= ☐ + ☐ + ☐
= ☐ + ☐
= ☐

10) 15920 − 4210

= ☐ + ☐ − ☐ + ☐ − ☐ + ☐ − ☐
= ☐ + ☐ + ☐ + ☐
= ☐ + ☐ + ☐
= ☐ + ☐
= ☐

Shortcut 21: Multiplying a number by 9

Rule
Multiplying by 9 is same as multiplying the number by 10 and then subtracting the original number from it. This is because 9 = 10 – 1.
Multiply the number by 10, which is same as putting a 'zero' next to it.
Subtract the original number from the number you get by multiplying by 10.
For subtracting, we will use shortcut #19.

Illustration

Calculate 78 X 9
Multiply by 10
78 X 10 = 780
Subtract the number from it:
780 – 78
Spilt 780 in 700 + 80
= 700 + 80 – 78
= 702

Examples:

Calculate 569 X 9
569 X 10 = 5690
5690 – 569
= 5000 + 690 – 569
= 5000 + 121
= 5121

Calculate 269 X 9
269 X 10 = 2690
= 2690 – 269
= 2000 + 690 – 269
= 2421

Calculate 23 X 9
23 X 10 = 230
= 230 – 23
= 200 + 30 – 23
= 200 + 7
= 207

Calculate 5898 X 9
5898 X 10 = 58980
58980 – 5898
= 50000 + 8000 – 5000 + 980 – 898
= 50000 + 3000 + 82
= 53082

Calculate 88 X 9
88 X 10 = 880
= 880 – 88
= 880 – 88
= 792

Calculate 982354 X 9
982354 X 10 = 9823540
9823540 – 982354
(we spilt 9823540 into 2 parts to simplify the subtraction)
= 8823540 + 1000000 – 982354
= 8823540 + 17646
= 8841186

Exercise:

1)   72 X 9
= 72 X 10 = [ ]
= [ ] − 72
= [ ] − 70 − 2
= [ ] − [ ]
= [ ]

2)   81 X 9
= 81 X 10 = [ ]
= [ ] − 81
= [ ] − [ ] − 1
= [ ] − [ ]
= [ ]

3)   113 X 9
= 113 X 10 = [ ]
= [ ] − 113
= [ ] − 100 − 10 − 3
= [ ] − 10 − 3
= [ ] − 3
= [ ]

4)   129 X 9
= 129 X 10 = [ ]
= [ ] − 129
= [ ] − 100 − 20 − 9
= [ ] − 20 − 9
= [ ] − 9
= [ ]

5) 983 X 9
=   983 X 10 = ☐
=   ☐ − 983
=   ☐ − 1000 + 17
=   ☐ + 17
=   ☐

6) 1123 X 9
=   1123 X 10 = ☐
=   ☐ − 1123
=   ☐ − 1000 − 100 − 23
=   ☐ − 100 − 23
=   ☐ − 23
=   ☐

7) 2091 X 9
=   2091 X 10 = ☐
=   ☐ − 2091
=   ☐ − ☐ − ☐ − ☐
=   ☐ − ☐ − ☐
=   ☐ − ☐
=   ☐

8)   2982   X   9
=   2982   X   10   =   ☐
=   ☐   −   2982
=   ☐ − ☐ − ☐ − ☐ − ☐
=   ☐ − ☐ − ☐ − ☐
=   ☐ − ☐ − ☐
=   ☐ − ☐
=   ☐

9)   1892   X   9
=   1892   X   10   −   ☐
=   ☐ − ☐
=   ☐ − ☐ − ☐ − ☐ − ☐
=   ☐ − ☐ − ☐ − ☐
=   ☐ − ☐ − ☐
=   ☐ − ☐
=   ☐

10)   12345   X   9
=   ☐ − ☐
=   ☐ − ☐ − ☐ − ☐ − ☐
=   ☐ − ☐ − ☐ − ☐ − ☐
=   ☐ − ☐ − ☐ − ☐
=   ☐ − ☐ − ☐
=   ☐ − ☐
=   ☐

Shortcut 22: Multiplying a number by 11

Rule
Write down the rightmost digit of the number being multiplied by 11 as it is.
Starting from the right most digit, add each digit to the digit on the left and put the result in the answer line.
Continue this way till you reach the left most digit (added the left most pair of numbers).
Put the left most digit in the answer line.

Think of multiplying by 11 as multiplying by (10 + 1).Then multiplying by 11 is same as multiplying by 10 and then adding the number to the product. But multiplying by 10 is same as putting a 'zero' at the end of the number – or shifting the number to the left by one space. This is same as adding each digit to the one on its left and putting that as the answer.

Illustration

Calculate 3122 X 11
Put the right most digit in the answer.
Answer: 2
Now starting from the right most digit, add each digit to the digit on its left and put the result in the answer line.
31<u>22</u>
You add 2 and 2 = 4 and put 4 in the answer line next to the 2 already there: 42.
3<u>12</u>2
Now add 2 and 1 = 3 and put 3 in the answer line: 342.
Add 1 and 3 = 4 and put the 4 there: 4342.
Now you have reached the left most digit. Put the left most digit in the answer line: 34342 – this is the answer.

3122 X 11 = 34342

Calculate 431 X 11
Right most digit as is 1.
Now we move leftwards and add each number to the one on its left:
1 + 3 = 4 (put in answer) 41
3 + 4 = 7 (put in answer) 741
We have reached the left most digit so put that down as is in the answer line:
Answer : 431 X 11 = 4741

There is one variation that you need to know and be careful about.

Illustration

Calculate 78 X 11

Put the right most digit in the answer line: 8.
Next, add the 8 to the digit on the left: 8 + 7 = 15.

The difference here is that our result is a two digit answer but we can put only one digit.
To handle this we:
Put the right digit in the answer line.
Carry over the remaining digit to the next step of the calculation.

So we put 5 in the answer line: 58 and carry over 1.
Now we have reached the left most digit. So we should put that in the answer line. Since we have a carry digit, we add that to the 7 to get 8.

1 + 7 = 8 and put 8 in the answer line to get the final answer: 858

78 X 11 = 858

The thing to remember is that as you keep going left after the right most digit, if the result is single digit you put that in the answer line. But if it is more than one digit answer, put the right most digit there and carry over the other digit to the next step.

Examples:
Calculate 43286 X 11
Put the right most digit in the answer: 6.

Add 6 to the digit on its left: 6 + 8 = 14.
Put 4 in the answer line and carry over 1.
(Answer: 46)

Now add 8 to the digit on its left: 8 + 2 = 10. Add the carry digit: 10 + 1 = 11.
Put 1 in the answer line and carry over 1.
(Answer: 146)

Add 2 with 3 (digit on its left):
2 + 3 = 5 + 1 (carry over) = 6
Put 6 in the answer line.
(Answer: 6146)

Add 3 with 4: 3 + 4 = 7.
No carry so put 7 in the answer line:
(Answer: 76146)

We have reached the left most digit so we put it in the answer line as is (again because of no carry digit) and that will give us the answer:

Answer: 43286 X 11 = 476146

Examples:
Calculate 7897 X 11

Right most digit into the answer: 7.
Next 7 + 9 = 16. 6 in the answer line and carry over 1.
(Answer: 67)

Next 9 + 8 = 17 + 1 (carry over) = 18. Put 8 in the answer and carry over 1.
(Answer: 867)

Next 8 + 7 = 15 + 1 (carry over) = 16. Put 6 in the answer and carry over 1.
(Answer: 6867)

We have reached the leftmost digit so put 7 in the answer line + 1 (carry over) = 7 + 1 = 8.

Answer: 7897 X 11 = 86867

Calculate 835 X 11

Right most digit into the answer: 5
Next 5 + 3 = 8.
(Answer: 85)

Next 3 + 8 = 11 Put 1 in the answer and carry over 1.
(Answer: 185)

We have reached the leftmost digit so put 8 in the answer line + 1 (carry over) = 9.

Answer: 835 X 11 = 9185

Calculate 298 X 11

Right most digit into the answer: 8.
Next 8 + 9 = 17
(Write 7 in answer line and carry over 1.)
(Answer: 78)

Next 9 + 2 = 11 + 1 (carry over)= 12.
(Write 2 in answer line and carry over 1.)
(Answer: 278)

We have reached the left most digit so put 2 in the answer line + 1 (carry over) = 3.

Answer : 298 X 11 = 3278

Examples:
Calculate 9563 X 11
Right most digit into the answer: 3.
Next 3 + 6 = 9.
(Answer: 93)

Next 6 + 5 = 11 Put 1 in the answer and carry over 1.
(Answer: 193)

Next 5 + 9 = 14 + 1 (carry over) = 15.
Put 5 in the answer and carry over 1.
(Answer: 5193)

We have reached the leftmost digit so put 9 in the answer line + 1 (carry over)= 10.

Answer: 9563 X 11 = 105193

Calculate 23153 X 11
Right most digit into the answer: 3.
Next 3 + 5 = 8.
(Answer: 83)

Next 5 + 1 = 6, put 6 in the answer.
(Answer: 683)

Next 1 + 3 = 4, put 4 in the answer.
(Answer: 4683)

Next 3 + 2 = 5, put 5 in the answer.
(Answer: 54683)

We have reached the leftmost digit so put 2 in the answer line.

Answer: 23153 X 11= 254683

Examples:
Calculate 56789 X 11

Right most digit into the answer: 9.
Next 9 + 8 = 17
Put 7 in the answer line and carry over 1.
(Answer: 79)

Next 8 + 7 = 15 + 1 (carry over) = 16.
Put 6 in the answer line and carry over 1.
(Answer: 679)

Next 7 + 6 = 13 + 1 (carry over) = 14.
Put 4 in the answer line and carry over 1.
(Answer: 4679)

Next 6 + 5 = 11 + 1 (carry over) = 12.
Put 2 in the answer line and carry over 1.
(Answer: 24679)

We have reached the leftmost digit so put 5 + 1 (carry over) = 6 in the answer line.

Answer: 56789 X 11 = 624679

Exercise:

1) 25 X 11 =

   Put right most digit into answer: .....

   Add 5 to the digit on it's left

   = 5 + [ ] = 7

   Put ... in answer line

   Answer: [ ]

   We have reached the left most digit so we put 2 in the answer line.

   25 X 11 = [ ]

---

2) 39 X 11 =

   Put right most digit into answer: .....

   Add 9 to the digit on it's left

   = 9 + [ ] = [ ]

   Put ... in answer line, carry 1

   Answer: [ ]

   We have reached the left most digit so we put 3 + ... (carry over) = in the answer line.

   39 X 11 = [ ]

3) 184 X 11 =

Put right most digit into answer: .....

Add 4 to the digit on it's left

= 4 + 8 = ☐

Put ... in answer line, carry over 1

Answer: ☐

Add 8 to the digit on it's left + 1 (carry over).

= 8 + 1 + 1 = ☐
        (carry over)

Put ... in answer line, carry over ...

Answer: ☐

We have reached the left most digit so we put 1 + ... (carry over) = ... in the answer line.

184 X 11 = ☐

---

4) 2084 X 11 =

Put right most digit into answer: .....

Add 4 to the digit on it's left

= ☐ + ☐ = ☐

Put ... in answer line, carry over 1.

Answer: ☐

Add 8 to the digit on it's left + ... (carry over).

= 8 + ☐ + 1 = ☐
              (carry over)

Put ... in answer line, carry over ...

Answer: ☐

Add '0' to the digit on it's left

= 0 + ☐ = 2

Answer: ☐

We have reached the left most digit so we put 2 in the answer line.

2084 X 11 = ☐

5) 5189 X 11 =

Put right most digit into answer: .....

Add 9 to the digit on it's left

= 9 + [ ] = [ ]

Put 7 in answer line, carry over ...

Answer: [ ]

Add 8 to the digit on it's left + ... (carry over).

= 8 + 1 + [ ] = [ ]
              (carry over)

Put 0 in answer line, carry over ...

Answer: [ ]

Add 1 to the digit on it's left + ... (carry over).

= 1 + [ ] + [ ] = [ ]
                  (carry over)

Put 7 in the answer line

Answer: [ ]

We have reached the left most digit so we put 5 in the answer line.

5189 X 11 = [ ]

---

6) 7895 X 11 = [ ]

---

7) 9265 X 11 = [ ]

---

8) 81375 X 11 = [ ]

---

9) 500989 X 11 = [ ]

---

10) 8993211 X 11 = [ ]

Shortcut 23: Multiplying a number by 25

Rule
The way to do a multiplication like this is to convert it to a simple division exercise. Since 25 X 4 is 100, multiplying by 25 is same as multiplying by 100 and then dividing by 4.

Multiplying by 100 is same as putting two 'zeroes' after the number.

Put two 'zeroes' after the given number.
Divide the number by 4.

Illustration

Calculate 32 X 25
This the same as:
$$\frac{32 \times 100}{4} = \frac{3200}{4}$$
= 800

32 X 25 = 800

Examples:

Calculate 76 X 25
= 76 X 100 = 7600
= $\frac{7600}{4}$
= 1900

Calculate 3275 X 25
= 3275 X 100 = 327500
= $\frac{327500}{4}$
= 81875

Calculate 12 X 25
= 12 X 100 = 1200
= $\frac{1200}{4}$
= 300

Calculate 38 X 25
= 38 X 100 = 3800
= $\frac{3800}{4}$
= 950

Calculate 989 X 25
= 989 X 100 = 98900
= $\frac{98900}{4}$
= 24725

Calculate 684 X 25
= 684 X 100 = 68400
= $\frac{68400}{4}$
= 17100

Calculate 5414 X 25
= 5414 X 100 = 541400
= $\frac{541400}{4}$
= 135350

Calculate 2822 X 25
= 2822 X 100 = 282200
= $\frac{282200}{4}$
= 70550

Exercise:

1)   14   X   25

= ☐ X 100 = ☐

= $\dfrac{\Box}{4}$

= ☐

2)   42   X   25

= ☐ X 100 = ☐

= $\dfrac{\Box}{4}$

= ☐

3)   98   X   25

= ☐ X 100 = ☐

= $\dfrac{\Box}{4}$

= ☐

4)   118   X   25

= ☐ X 100 = ☐

= $\dfrac{\Box}{4}$

= ☐

5)   139   X   25

= ☐ X 100 = ☐

= $\dfrac{\Box}{4}$

= ☐

6) 167 X 25
= ☐ X 100 = ☐
= ☐/4
= ☐

7) 485 X 25
= ☐ X ☐ = ☐
= ☐/☐
= ☐

8) 697 X 25
= ☐ X ☐ = ☐
= ☐/☐
= ☐

9) 1824 X 25
= ☐ X ☐ = ☐
= ☐/☐
= ☐

10) 2980 X 25
= ☐ X ☐ = ☐
= ☐/☐
= ☐

Shortcut 24: Squaring numbers that end in 5

Rule
Look at the part of the number other than the 5.
Multiply that number by one more than itself.
Put 25 after the result you get in the step above.

Illustration

$35^2 =$
The part other than the 5 is 3.
We have to multiply 3 by one more than itself.
Which means multiply 3 by 4 (4 = 3 + 1).
3 x 4 = 12
Put 25 after this to get: 1225
$35^2 = 1225$

Examples:

$15^2 =$
The part other than 5 = 1
1 x 2 = 2
Put 25 after it:
$15^2 = 225$

$25^2 =$
The part other than 5 = 2
2 x 3 = 6
Put 25 after it:
$25^2 = 625$

$65^2 =$
The part other than 5 = 6
6 x 7 = 42
Put 25 after it:
$65^2 = 4225$

$95^2 =$
9 x 10 = 90
$95^2 = 9025$

$205^2 =$
20 x 21 = 420
$205^2 = 42025$

$105^2 =$
10 x 11 = 110
$105^2 = 11025$

Exercise:

1) $45^2 =$

   $4 \times 5 = \boxed{}$

   Put 25 after it to get the answer:

   $45^2 = \boxed{}$

---

2) $75^2 =$

   $7 \times 8 = \boxed{}$

   $75^2 = \boxed{}$

---

3) $85^2 =$

   $\boxed{} \times \boxed{} = \boxed{}$

   $85^2 = \boxed{}$

---

4) $125^2 =$

   $\boxed{} \times \boxed{} = \boxed{}$

   $125^2 = \boxed{}$

---

5) $155^2 =$

   $\boxed{} \times \boxed{} = \boxed{}$

   $155^2 = \boxed{}$

6) $115^2 =$

   ☐ × ☐ = ☐

   $115^2 =$ ☐

---

7) $175^2 =$

   ☐ × ☐ = ☐

   $175^2 =$ ☐

---

8) $305^2 =$

   ☐ × ☐ = ☐

   $305^2 =$ ☐

---

9) $505^2 =$

   ☐ × ☐ = ☐

   $505^2 =$ ☐

---

10) $1005^2 =$

    ☐ × ☐ = ☐

    $1005^2 =$ ☐

Shortcut 25: Multiplying two number with same first digit and last digits that add to 10

Rule
Example: 28 X 22 or 11 X 19 etc.
Here the last digits add up to 10 (8 + 2 = 10 and 1 + 9 = 10. And the other part of the number is the same).

Look at the part that is same in both the numbers (other than the parts that add up to 10).
Multiply this with one more than itself.
Multiply the parts of the two numbers that add up to 10 and put the result after the answer you get above. If this is a single digit result, put a 'zero' before it.

Illustration

Suppose you have to calculate 27 X 23
The common part of the two numbers is 2.
Multiply this with 1 more than itself. So you do:
2 X 3 = 6
Multiply the parts that add up to 10.
7 X 3 = 21 and put it after the result you got above.
Answer: 27 x 23 = 621

Examples:

Calculate 72 X 78
Multiply 7 with one more than itself.
7 X 8 = 56
Multiply the parts that add up to 10 and put that after 56.
2 X 8 = 16
So 72 X 78 = 5616

Calculate 38 X 32
Multiply 3 with one more than itself.
3 X 4 = 12
Multiply the parts that add up to 10 and put that after 12.
8 X 2 = 16
so 38 X 32 = 1216

Calculate 91 X 99
Multiply 9 with one more than itself.
9 X 10 = 90
Multiply the parts that add up to 10 and put that after 90.
9 X 1 = 9 (put a '0' before 9 as it is single digit.)
So 91 X 99 = 9009

Calculate 57 X 53
Multiply 5 with one more than itself.
5 X 6 = 30
Multiply the parts that add up to 10 and put that after 30.
7 X 3 = 21
So 57 X 53 = 3021

Calculate 35 X 35
Multiply 3 with one more than itself.
3 X 4 = 12
Multiply the parts that add up to 10 and put that after 12.
5 X 5 = 25
So 35 X 35 = 1225

Calculate 24 X 26
Multiply 2 with one more than itself.
2 X 3 = 6
Multiply the parts that add up to 10 and put that after 6.
6 X 4 = 24
So 24 X 26 = 624

Examples:

Calculate 114 X 116
Multiply 11 with one more than itself.
11 X 12 = 132
Multiply the parts that add up to 10 and put that after 132.
4 X 6 = 24.
So 114 X 116 = 13224

Calculate 243 X 247
Multiply 24 with one more than itself.
24 X 25 = 600
Multiply the parts that add up to 10 and put that after 600.
3 X 7 = 21
So 243 X 247 = 60021

Calculate 138 X 132
Multiply 13 with one more than itself.
13 X 14 = 182
Multiply the parts that add up to 10 and put that after 182.
8 X 2 = 16
So 138 X 132 = 18216

Calculate 5008 X 5002
Multiply 500 with one more than itself.
500 X 501 = 250500
Multiply the parts that add up to 10 and put that after 250500.
8 X 2 = 16
So 5008 X 5002 = 25050016

Exercise:

1) 28 X 22

= ☐ X ☐ = ☐

= 8 X 2 = ☐

   28 X 22 = ☐

---

2) 11 X 19

= 1 X 2 = ☐

= 1 X 9 = ☐   (remember to put a 'zero' before it as it is single digit.)

   11 X 19 = ☐

---

3) 25 X 25

= 2 X ☐ = ☐

= ☐ X 5 = ☐

   25 X 25 = ☐

---

4) 64 X 66

= ☐ X ☐ = ☐

= ☐ X ☐ = ☐

   64 X 64 = ☐

---

5) 87 X 83

= ☐ X ☐ = ☐

= ☐ X ☐ = ☐

   87 X 83 = ☐

6)   112   X   118
=   11   X   [   ]   =   [   ]
=   2   X   8   =   [   ]
     112   X   118   =   [   ]

---

7)   127   X   123
=   [   ]   X   [   ]   =   [   ]
=   [   ]   X   [   ]   =   [   ]
     123   X   127   =   [   ]

---

8)   89   X   81
=   [   ]   X   [   ]   =   [   ]
=   [   ]   X   [   ]   =   [   ]
     89   X   81   =   [   ]

---

9)   47   X   43
=   [   ]   X   [   ]   =   [   ]
=   [   ]   X   [   ]   =   [   ]
     47   X   43   =   [   ]

---

10) 109   X   101
=   [   ]   X   [   ]   =   [   ]
=   [   ]   X   [   ]   =   [   ]
     109   X   101   =   [   ]

Shortcut 26: Squaring numbers near 50

Rule
Subtract 50 from the number.
Add the result to 25.
Square the difference of the number from 50 and put that after the result you got in the step above. If this square is a single digit number, then put a '0' before it to make it a two digit number.
If this square is a three digit number, then keep the right two digits and carry over the remaining digit to the left side.
If this square is a single digit number, put a zero before it.

Illustration

$56^2 =$
$56 - 50 = 6$ (Difference of the number from 50.)
$25 + 6 = 31$ (Add the difference to 25.)
$6^2 = 36$ (Square the difference from 50.)
$56^2 = 3136$ (Put this result at the end of the answer you got in step 2.)

$46^2 =$
$46 - 50 = -4$ (Difference of the number from 50.)
$25 + (-4) = 21$ (Because 46 is less than 50, we subtract 4 from 25.)
$4^2 = 16$ (Square the difference from 50.)
$46^2 = 2116$ (Put this result at the end of the answer you got in step 2.)

Examples:

$54^2 =$
$54 - 50 = 4$
$25 + 4 = 29$
$4^2 = 16$
$54^2 = 2916$

$59^2 =$
$59 - 50 = 9$
$25 + 9 = 34$
$9^2 = 81$
$59^2 = 3481$

$52^2 =$
$52 - 50 = 2$
$25 + 2 = 27$
$2^2 = 04$ (Note: This should be a two digit result. If it is a single digit, put a '0' before it.)
$52^2 = 2704$

$49^2 =$
$49 - 50 = -1$
$25 + (-1) = 24$ (Since the number to be squared is less than 50, then we should reduce that much from 25.)
$(-1)^2 = 01$ (This should be a two digit result. If it is a single digit, put a '0' before it.)
$49^2 = 2401$

$61^2 =$
$61 - 50 = 11$
$25 + 11 = 36$
$11^2 = 121$ [This should be a two digit result. But we have a three digit answer. From the 121, we will keep the right most two digits (21) and the third digit becomes a carry over to 36.]
$61^2 = 36_121$

$61^2 = 3721$

Exercise:

1) $57^2 =$

   $57 - 50 = \boxed{}$

   $25 + \boxed{} = \boxed{}$

   $\boxed{}^2 = \boxed{}$

   $57^2 = \boxed{}$

---

2) $51^2 =$

   $\boxed{} - \boxed{} = \boxed{}$

   $25 + \boxed{} = \boxed{}$

   $\boxed{}^2 = \boxed{}$

   $51^2 = \boxed{}$

   (Note: remember to put a '0' because the square is a single digit number.)

---

3) $48^2 =$

   $\boxed{} - \boxed{} = \boxed{}$

   $25 + \boxed{} = \boxed{}$

   $\boxed{}^2 = \boxed{}$

   $48^2 = \boxed{}$

   (Note: you have to subtract from 25 because 48 is less than 50.)

4) $55^2 =$

☐ − ☐ = ☐

25 + ☐ = ☐

☐^{2} = ☐

55^2 = ☐

---

5) $62^2 =$

☐ − ☐ = ☐

☐ + ☐ = ☐

☐^{2} = ☐

62^2 = ☐

(The square is a three digit number. Remember to keep the right two digits and carry one the left digit.)

---

6) $53^2 =$

☐ − ☐ = ☐

☐ + ☐ = ☐

☐^{2} = ☐

53^2 = ☐

---

7) $42^2 =$

☐ − ☐ = ☐

☐ + ☐ = ☐

☐^{2} = ☐

42^2 = ☐

Shortcut 27: Squaring numbers that end in '0'

Rule
Count the number of 'zeroes' in the number being squared.
Ignore the 'zeroes', take the remaining part of the number and square it.
Put the <u>double the number of 'zeroes'</u> after the result above.

Illustration

$30^2 =$
Number of zeroes = 1
Ignoring the zero we have 3
$3^2 = 9$
Since the number being squared had one 'zero', we put two 'zeroes' after 3^2 to get answer:
$30^2 = 900$

Examples:

$50^2 =$
Number of zeroes = 1
$5^2 = 25$
$50^2 = 2500$ (put double the number of 'zeroes' after it.)

$70^2 =$
Number of zeroes = 1
$7^2 = 49$
$70^2 = 4900$

$120^2 =$
Number of zeroes = 1
$12^2 = 144$
$120^2 = 14400$

$190^2 =$
Number of zeroes = 1
$19^2 = 361$
$190^2 = 36100$

$200^2 =$
Number of zeroes = 2
$2^2 = 4$
$200^2 = 40000$
(We put four 'zeroes' after the result because the original number had two 'zeroes'.)

$1000^2 =$
Number of zeroes = 3
$1^2 = 1$
$1000^2 = 1000000$ (6 'zeroes')

Exercise:

1) $20^2 =$

Number of zeroes = [ ]

2^2 = [ ]

20^2 = [ ]

(Square the part other than the 'zero' and put double the number of 'zeroes' after that.)

---

2) $40^2 =$

Number of zeroes = [ ]

4^2 = [ ]

40^2 = [ ]

(Square the part other than the 'zero' and put double the number of 'zeroes' after that.)

---

3) $60^2 =$

Number of zeroes = [ ]

6^2 = [ ]

60^2 = [ ]

(Square the part other than the 'zero' and put double the number of 'zeroes' after that.)

---

4) $80^2 =$

Number of zeroes = [ ]

8^2 = [ ]

80^2 = [ ]

5) 90^2 =

Number of zeroes = ☐

9^2 = ☐

90^2 = ☐ (Put double the number of 'zeroes' after 9^2.)

---

6) 120^2 =

Number of zeroes = ☐

12^2 = ☐

120^2 = ☐

---

7) 150^2 =

Number of zeroes = ☐

15^2 = ☐

150^2 = ☐

---

8) 210^2 =

Number of zeroes = ☐

21^2 = ☐

210^2 = ☐

---

9) 250^2 =

Number of zeroes = ☐

25^2 = ☐

250^2 = ☐

10) $300^2 =$

Number of zeroes = [ ]

30^2 = [ ]

300^2 = [ ]

Shortcut 28: Squaring numbers that end in '1'

Rule
Square the number 1 less than this number (refer to the previous shortcut).
To that result, add the number being squared and the number 1 less than it.

Illustration

$51^2 =$
$50^2 = 2500$ (Square of the number 1 less.)
$51^2 = 2500 + 50 + 51$ (To the square add the number being squared and 1 less than it.)

$51^2 = 2601$

Examples:

$61^2 =$
$60^2 = 3600$
$61^2 = 3600 + 60 + 61$
$61^2 = 3721$

$71^2 =$
$70^2 = 4900$
$71^2 = 4900 + 70 + 71$
$ = 4900 + 141$
$ = 5041$

$121^2 =$
$120^2 = 14400$
$121^2 = 14400 + 120 + 121$
$ = 14400 + 241$
$ = 14641$

$191^2 =$
$190^2 = 36100$
$191^2 = 36100 + 190 + 191$
$ = 36100 + 381$
$ = 36481$

$201^2 =$
$200^2 = 40000$
$201^2 = 40000 + 200 + 201$
$ = 40000 + 401$
$ = 40401$

$1001^2 =$
$1000^2 = 1000000$
$1001^2 = 1000000 + 1000 + 1001$
$ = 1000000 + 2001$
$ = 1002001$

Exercise:

1) $21^2 =$

   $20^2 = \boxed{}$

   $21^2 = \boxed{} + 20 + 21$

   $21^2 = \boxed{}$

2) $31^2 =$

   $30^2 = \boxed{}$

   $31^2 = \boxed{} + 30 + 31$

   $31^2 = \boxed{}$

3) $41^2 =$

   $40^2 = \boxed{}$

   $41^2 = \boxed{} + 40 + 41$

   $41^2 = \boxed{}$

4) $51^2 =$

   $50^2 = \boxed{}$

   $51^2 = \boxed{} + 50 + 51$

   $51^2 = \boxed{}$

5) $81^2 =$

   $80^2 = \boxed{}$

   $81^2 = \boxed{} + 80 + 81$

   $81^2 = \boxed{}$

6) $151^2 =$

   $150^2 = \boxed{}$

   $151^2 = \boxed{} + 150 + 151$

   $151^2 = \boxed{}$

7) $111^2 =$

   $110^2 = \boxed{}$

   $111^2 = \boxed{} + \boxed{} + 111$

   $111^2 = \boxed{}$

8) $131^2 =$

   $130^2 = \boxed{}$

   $131^2 = \boxed{} + \boxed{} + \boxed{}$

   $131^2 = \boxed{}$

9) $181^2 =$

$180^2 = \boxed{}$

$181^2 = \boxed{} + \boxed{} +$
$\boxed{}$

$181^2 = \boxed{}$

10) $291^2 =$

$290^2 = \boxed{}$

$291^2 = \boxed{} + \boxed{} +$
$\boxed{}$

$291^2 = \boxed{}$

Shortcut 29: Multiplying numbers that are equidistant from a multiple of 10

Rule
Let the number from which both these numbers are equidistant be "central" number.

1. Square the central number.
2. Calculate the difference of these numbers from the central number.
3. Square this number.
4. Subtract the number you got in step 3 from the number you got in step 1.

Illustration
Calculate 46 X 54
Central number is 50 as it is 4 away from both the numbers we want to multiply.
54 – 50 = 4
4 X 4 = 16
50 X 50 = 2500
2500 – 16 = 2484
46 X 54 = 2484

Examples:
Calculate 77 X 83
Central number = 80
Difference from central number = 3
80 X 80 = 6400
3 X 3 = 9
6400 – 9 = 6391

77 X 83 = 6391

Calculate 22 X 38
Central number = 30
Difference from central number = 8
30 X 30 = 900
8 X 8 = 64
900 – 64 = 836

22 X 38 = 836

Calculate 88 X 92
Central number = 90
Difference from central number = 2
90 X 90 = 8100
2 X 2 = 4
8100 – 4 = 8000 + 100 – 4 = 8096

88 X 92 = 8096

Calculate 114 X 106
Central number = 110
Difference from central number = 4
110 X 110 = 12100
4 X 4 = 16
12100 – 16 = 12000 + 100 – 16
= 12084

114 X 106 = 12084

Calculate 27 X 33
Central number = 30
Difference from central number = 3
30 X 30 = 900
3 X 3 = 9
900 – 9 = 891

27 X 33 = 891

Calculate 83 X 97
Central number = 90
Difference from central number = 7
90 X 90 = 8100
7 X 7 = 49
= 8100 – 49 = 8000 + 100 – 49
= 8051
83 X 97 = 8051

Examples:
Calculate 34 X 46
Central number = 40
Difference from central number = 6
40 X 40 = 1600
6 X 6 = 36
1600 − 36 = 1000 + 600 − 36
= 1564

34 X 46 = 1564

Calculate 96 X 104
Central number = 100
Difference from central number = 4
100 X 100 = 10000
4 X 4 = 16
10000 − 16 = 9984

96 X 104 = 9984

Exercise:

1) 43 X 57
    Central Number = 50
    Difference from central number = ☐
    = 50 X 50 = ☐
    = 7 X 7 = ☐
    = ☐ − ☐
    = ☐

2) 87 X 93
    Central Number = 90
    Difference from central number = ☐
    = ☐ X ☐ = ☐
    = ☐ X ☐ = ☐
    = ☐ − ☐
    = ☐

3) 22 X 18
    Central Number = ☐
    Difference from central number = 2
    = ☐ X ☐ = ☐
    = ☐ X ☐ = ☐
    = ☐ − ☐
    = ☐

4)   36   X   44
     Central Number = ☐
Difference from central number = 4

= ☐ X ☐ = ☐
= ☐ X ☐ = ☐
= ☐ − ☐
= ☐

---

5)   45   X   55
     Central Number = ☐
Difference from central number = ☐

= ☐ X ☐ = ☐
= ☐ X ☐ = ☐
= ☐ − ☐
= ☐

---

6)   292   X   308
     Central Number = ☐
Difference from central number = ☐

= ☐ X ☐ = ☐
= ☐ X ☐ = ☐
= ☐ − ☐
= ☐

7)   173   X   167
   Central Number = 170
   Difference from central number = 3
   = [170] X [170] = [28900]
   = [3] X [3] = [9]
   = [28900] − [9]
   = [28891]

8)   493   X   507
   Central Number = 500
   Difference from central number = 7
   = [500] X [500] = [250000]
   = [7] X [7] = [49]
   = [250000] − [49]
   = [249951]

9)   362   X   358
   Central Number = [360]
   Difference from central number = [2]
   = [360] X [360] = [129600]
   = [2] X [2] = [4]
   = [129600] − [4]
   = [129596]

10) 608 X 592

Central Number = ☐

Difference from central number = ☐

= ☐ X ☐ = ☐

= ☐ X ☐ = ☐

= ☐ − ☐

= ☐

Shortcut 30: Adding a series of consecutive numbers from 'a' to 'b'

Note: It is important for the method below to work that 'a' be greater than '0', i.e. 'a' should be a positive number.

Rule:
If you want to add all numbers in the form:
$a + (a + 1) + (a + 2) + \ldots + b$

1. Square the last number and the first number.
2. From the square of the last number, subtract the square of the first number.
3. To the difference above, add the first and the last number.
4. Divide the total sum by 2.

Examples:

What is 5 + 6 + 7 ... + 14 + 15
$15^2 = 225$
$5^2 = 25$
Subtract the square of first number from the square of last number.
$15^2 - 5^2 = 225 - 25 = 200$
To the difference add the first number and the last number.
$200 + 15 + 5 = 220$
Divide the total by 2
$= 220 \div 2$
$= 110$

What is 7 + 8 + 9 +... + 19 + 20
$20^2 = 400$
$7^2 = 49$
$20^2 - 7^2 = 400 - 49 = 351$
$351 + 20 + 7 = 378$
$= 378 \div 2$
$= 189$

What is 10 + 11 + 12 ... + 24 + 25
$25^2 = 625$
$10^2 = 100$
Subtract the square of first number from the square of the last number.
$25^2 - 10^2 = 625 - 100 = 525$
To the difference above, add first number and the last number.
$525 + 25 + 10 = 560$
Divide the total by 2
$= 560 \div 2$
$= 280$

What is 11 + 12 + 13 ... + 18 + 19
$19^2 = 361$
$11^2 = 121$
$19^2 - 11^2 = 361 - 121 = 240$
$240 + 19 + 11 = 270$
$= 270 \div 2$
$= 135$

What is 3 + 4 + 5 ... + 11 + 12
$12^2 = 144$
$3^2 = 9$
$12^2 - 3^2 = 144 - 9 = 135$
$135 + 12 + 3 = 150$
$= 150 \div 2$
$= 75$

What is 5 + 6 + 7 +... + 44 + 45
$45^2 = 2025$
$5^2 = 25$
$45^2 - 5^2 = 2025 - 25 = 2000$
$2000 + 45 + 5 = 2050$
$= 2050 \div 2$
$= 1025$

Exercise:

1) $\quad 9+10+11\ldots\ldots+19+20$

$= \quad 20^2 \quad = \quad \boxed{}$

$= \quad 9^2 \quad = \quad \boxed{}$

$= \quad \boxed{} - \boxed{} = \boxed{}$

$= \quad \boxed{} + 20 + 9 = \boxed{}$

$= \quad \dfrac{\boxed{}}{2}$

$= \quad \boxed{}$

---

2) $\quad 15+16+17\ldots\ldots+29+30$

$= \quad 30^2 \quad = \quad \boxed{}$

$= \quad 15^2 \quad = \quad \boxed{}$

$= \quad \boxed{} - \boxed{} = \boxed{}$

$= \quad \boxed{} + 30 + 15 = \boxed{}$

$= \quad \dfrac{\boxed{}}{2}$

$= \quad \boxed{}$

---

3) $\quad 5+6+7+\ldots\ldots+19+20$

$= \quad 20^2 \quad = \quad \boxed{}$

$= \quad 5^2 \quad = \quad \boxed{}$

$= \quad \boxed{} - \boxed{} = \boxed{}$

$= \quad \boxed{} + \boxed{} + \boxed{} = \boxed{}$

$= \quad \dfrac{\boxed{}}{2}$

$= \quad \boxed{}$

4) $\quad 10+11+12+\ldots+29+30$

$= \boxed{}^2 = \boxed{}$

$= \boxed{}^2 = \boxed{}$

$= \boxed{} - \boxed{} = \boxed{}$

$= \boxed{} + \boxed{} + \boxed{} = \boxed{}$

$= \dfrac{\boxed{}}{2}$

$= \boxed{}$

---

5) $\quad 7+8+9+\ldots+34+35$

$= \boxed{}^2 = \boxed{}$

$= \boxed{}^2 = \boxed{}$

$= \boxed{} - \boxed{} = \boxed{}$

$= \boxed{} + \boxed{} + \boxed{} = \boxed{}$

$= \dfrac{\boxed{}}{2}$

$= \boxed{}$

---

6) $\quad 20+21+22+\ldots+34+35$

$= \boxed{}^2 = \boxed{}$

$= \boxed{}^2 = \boxed{}$

$= \boxed{} - \boxed{} = \boxed{}$

$= \boxed{} + \boxed{} + \boxed{} = \boxed{}$

$= \dfrac{\boxed{}}{2}$

$= \boxed{}$

7) $25+26+27+\ldots+39+40$

$= \boxed{}^2 = \boxed{}$

$= \boxed{}^2 = \boxed{}$

$= \boxed{} - \boxed{} = \boxed{}$

$= \boxed{} + \boxed{} + \boxed{} = \boxed{}$

$= \dfrac{\boxed{}}{2}$

$= \boxed{}$

---

8) $10+16+17+\ldots+34+35$

$= \boxed{}^2 = \boxed{}$

$= \boxed{}^2 = \boxed{}$

$= \boxed{} - \boxed{} = \boxed{}$

$= \boxed{} + \boxed{} + \boxed{} = \boxed{}$

$= \dfrac{\boxed{}}{2}$

$= \boxed{}$

---

9) $2+3+4+\ldots+39+40$

$= \boxed{}^2 = \boxed{}$

$= \boxed{}^2 = \boxed{}$

$= \boxed{} - \boxed{} = \boxed{}$

$= \boxed{} + \boxed{} + \boxed{} = \boxed{}$

$= \dfrac{\boxed{}}{\boxed{}}$

$= \boxed{}$

10) $10+11+12+\ldots+19+20$

= $\boxed{}^2 = \boxed{}$

= $\boxed{}^2 = \boxed{}$

= $\boxed{} - \boxed{} = \boxed{}$

= $\boxed{} + \boxed{} + \boxed{} = \boxed{}$

= $\dfrac{\boxed{}}{\boxed{}}$

= $\boxed{}$

Shortcut 31: Adding a series of numbers from 'a' to 'b'

This is a series of numbers where there is a <u>constant difference</u> between consecutive numbers. Meaning each number is equal to the previous number plus a <u>constant difference.</u>

Example **2 + 5 + 8 + 11 + ...+ 32**
Here the difference between consecutive numbers is 3. So if we add 3 to any number, it will give us the next number.

To find the sum, we use three numbers in the calculation:
First number
Last number and
Constant Difference

Rule
Calculate the number of terms in the series of number

$$\text{number of term} = \left[\frac{(\text{last number} - \text{first number})}{\text{constant difference}} + 1 \right]$$

$$\text{sum} = (\text{number of term}) \times \left[\frac{\text{last number} + \text{first number}}{2} \right]$$

Illustration

What is 2 + 5 + 8 + 11 + ...+ 32
First number = 2
Last number = 32
Constant Difference = 3 (difference between consecutive numbers is 3 = constant difference.)

Number of term = $\left[\frac{32 - 2}{3} + 1 \right]$ = 11

Sum = 11 X $\left[\frac{32 + 2}{2} \right]$ = 11 X 17 = 187

Examples:

What is 4 + 9 + 14 + 19 +...+ 94 + 99

First number = 4
Last number = 99
Constant Difference = 5
Last number − first number = 99 − 4 = 95

Number of term = $\left[\dfrac{99-4}{5} + 1\right]$ = 20

Sum of term = 20 × $\left[\dfrac{99+4}{2}\right]$ = 20 × $\dfrac{103}{2}$

= 1030

What is 5 + 10 + 15 + 20 + ... + 120 + 125

First number = 5
Last number = 125
Constant Difference = 5

Number of term = $\left[\dfrac{125-5}{5} + 1\right]$ = $\dfrac{120}{5} + 1$

= 25

Sum of term = 25 × $\left[\dfrac{125+5}{2}\right]$ = 25 × $\dfrac{130}{2}$

= 25 × 65
= 1625

What is 12 + 21 + 30 + 39 + ... + 237 + 246

First number = 12
Last number = 246
Constant Difference = 9

Number of term = $\left[\dfrac{246-12}{9} + 1\right]$ = 27

Sum of term = 27 × $\left[\dfrac{246+12}{2}\right]$ = 27 × 129

= 3483

Examples:

What is 22 + 30 + 38 + 46 + ... + 78 + 86
First number = 22
Last number = 86
Constant Difference = 8

Number of term = $\left[\dfrac{86 - 22}{8} + 1\right]$ = 9

Sum of term = $9 \times \left[\dfrac{86 + 22}{2}\right]$ = $9 \times \dfrac{108}{2}$

$= 9 \times 54$
$= 486$

What is 1 + 5 + 9 + 13 + ... + 33 + 37
First number = 1
Last number = 37
Constant Difference = 4

Number of term = $\left[\dfrac{37 - 1}{4} + 1\right]$ = 10

Sum of term = $10 \times \left[\dfrac{37 + 1}{2}\right]$ = 10×19

$= 190$

What is 2 + 9 + 16 + 23 + ... + 51 + 58
First number = 2
Last number = 58
Constant Difference = 7

Number of term = $\left[\dfrac{58 - 2}{7} + 1\right]$ = 9

Sum of term = $9 \times \left[\dfrac{58 + 2}{2}\right]$ = $9 \times \dfrac{60}{2}$

$= 9 \times 30$
$= 270$

Exercise:

1) $1 + 3 + 5 + \ldots + 22 + 23$

   Number of term $= \left[\dfrac{23 - 1}{2} + 1 \right] = \boxed{}$

   Sum $= \boxed{} \times \left[\dfrac{23 + 1}{2} \right]$

   $= \boxed{} \times \boxed{}$

   $= \boxed{}$

---

2) $2 + 4 + 6 + \ldots + 30 + 32$

   Number of term $= \left[\dfrac{\boxed{} - \boxed{}}{2} + 1 \right] = \boxed{}$

   Sum $= \boxed{} \times \left[\dfrac{\boxed{} + \boxed{}}{2} \right]$

   $= \boxed{} \times \boxed{}$

   $= \boxed{}$

---

3) $3 + 6 + 9 + \ldots + 42 + 45$

   Number of term $= \left[\dfrac{\boxed{} - \boxed{}}{3} + 1 \right] = \boxed{}$

   Sum $= \boxed{} \times \left[\dfrac{\boxed{} + \boxed{}}{2} \right]$

   $= \boxed{} \times \boxed{}$

   $= \boxed{}$

4) 5 + 8 + 11 + ... + 26 + 29

Number of term = $\left[\dfrac{\Box - \Box}{\Box} + 1\right]$ = $\Box$

Sum = $\Box$ × $\left[\dfrac{\Box + \Box}{2}\right]$

= $\Box$ × $\Box$

= $\Box$

---

5) 3 + 8 + 13 + ... + 38 + 43

Number of term = $\left[\dfrac{\Box - \Box}{\Box} + 1\right]$ = $\Box$

Sum = $\Box$ × $\left[\dfrac{\Box + \Box}{2}\right]$

= $\Box$ × $\Box$

= $\Box$

---

6) 1 + 7 + 13 + ... + 43 + 49

Number of term = $\left[\dfrac{\Box - \Box}{\Box} + 1\right]$ = $\Box$

Sum = $\Box$ × $\left[\dfrac{\Box + \Box}{2}\right]$

= $\Box$ × $\Box$

= $\Box$

7) 5 + 9 + 13 + ... + 49 + 53

Number of term = $\left[\dfrac{\boxed{} - \boxed{}}{\boxed{}} + 1 \right]$ = $\boxed{}$

Sum = $\boxed{}$ × $\left[\dfrac{\boxed{} + \boxed{}}{2} \right]$

= $\boxed{}$ × $\boxed{}$

= $\boxed{}$

---

8) 1 + 10 + 19 + ... + 53 + 64

Number of term = $\left[\dfrac{\boxed{} - \boxed{}}{\boxed{}} + 1 \right]$ = $\boxed{}$

Sum = $\boxed{}$ × $\left[\dfrac{\boxed{} + \boxed{}}{2} \right]$

= $\boxed{}$ × $\boxed{}$

= $\boxed{}$

---

9) 5 + 10 + 15 + ... + 55 + 60

Number of term = $\left[\dfrac{\boxed{} - \boxed{}}{\boxed{}} + 1 \right]$ = $\boxed{}$

Sum = $\boxed{}$ × $\left[\dfrac{\boxed{} + \boxed{}}{2} \right]$

= $\boxed{}$ × $\boxed{}$

= $\boxed{}$

10) 11 + 21 + 31 + ... + 71 + 81

Number of term = $\left[\dfrac{\boxed{} - \boxed{}}{\boxed{}} + 1\right]$ = $\boxed{}$

Sum = $\boxed{}$ × $\left[\dfrac{\boxed{} + \boxed{}}{2}\right]$

= $\boxed{}$ × $\boxed{}$

= $\boxed{}$

Shortcut 32: Multiplying numbers with same initial digit and other digits of both numbers add up to 100

Rule

1. Multiply the last two digit of both numbers that add up to 100. This will form the last four digits of the answer.
2. Now multiply the hundred's place digit (which is same in both numbers) with one more than itself. And put this at the beginning of the result in the step 1.

Illustration

291 X 209

1. Multiply the last two digits that add up to 100 i.e. 91 X 09 = 819. So last four digits of the answer are 0819. Put '0' before as it has to be four digits number.
2. Now multiply the digit at hundred's place one more then itself i.e. 2 X 3 = 6.
3. Put 6 before 0819 to get 291 X 209 = 60819.

Examples:

396 X 304

1. Multiply the last two digits that add up to 100 i.e. 96 X 04 = 384. Last four digits = 0384 (add '0' before 384 as the answer of this step has to be four digits number).
2. The digit at hundred's place one more than itself 3 X 4 = 12.
3. Put 12 before 0384 to get 396 X 304 = 120384.

246 X 254

1. 46 X 54
   $= 50^2 - 4^2$ (refer shortcut 26)
   = 2500 – 16
   = 2484
   Since this is already four digits, so we won't put a '0' before it.
2. 2 X 3 = 6
3. 246 X 254 = 62484

989 X 911

1. 89 X 11
   = 80 X 11 + 9 X 11
   = 880 + 99
   = 0979
2. 9 X 10 = 90
3. 989 X 911 = 900979

Examples:
721 X 779
1. 21 X 79
    = 20 X 79 + 1 X 79
    = 1580 + 79
    = 1659
2. 7 X 8 = 56
3. 721 X 779 = 561659

880 X 820
1. 80 X 20 = 1600
2. 8 X 9 = 72
3. 880 X 820 = 721600

540 X 560
1. 40 X 60 = 2400
2. 5 X 6 = 30
3. 540 X 560 = 302400

Exercise:

1) 493 X 407

    93 X 7 = [ ]

    4 X 5 = 20

    493 X 407 = [ ]

---

2) 879 X 821

    79 X 21 = [ ]

    8 X 9 = [ ]

    879 X 821 = [ ]

---

3) 557 X 543

    57 X 43 = [ ]

    5 X 6 = [ ]

    557 X 543 = [ ]

---

4) 669 X 631

    69 X 31 = [ ]

    6 X 7 = [ ]

    669 X 631 = [ ]

5)   752   X   748
     52   X   [   ]
     7   X   [   ]   =   [   ]
     752   X   748   =   [   ]

---

6)   985   X   915
     [   ]   X   15
     9   X   10   =   [   ]
     985   X   915   =   [   ]

---

7)   991   X   909
     [   ]   X   [   ]
     [   ]   X   [   ]   =   [   ]
     991   X   909   =   [   ]

---

8)   372   X   328
     [   ]   X   [   ]
     [   ]   X   [   ]   =   [   ]
     372   X   328   =   [   ]

9) 462   X   438
    [   ] X [   ]
    [   ] X [   ] = [   ]
    462   X   438   = [   ]

10) 770   X   730
    [   ] X [   ]
    [   ] X [   ] = [   ]
    770   X   730   = [   ]

Shortcut 33: Dividing numbers by 9

Rule
We will do this method first for two-digit numbers then expand it to larger numbers.

1. Write the left digit as the quotient and the sum of both the digits as the remainder.
2. If the remainder is more than 9, <u>then see how many times 9 divides into the remainder.</u> Add that number to the quotient you got above and the net remainder as the remainder.

Illustration

31 ÷ 9 =
Left digit is 3, that will be the quotient and sum of the digits = 3 + 1 = 4, this will be the remainder.

31 ÷ 9 = 3 (quotient) and remainder = 4.

61 ÷ 9 =
Left digit is 6 (quotient).
Sum of the digits = 6 + 1 = 7 (remainder).

61 ÷ 9 = 6 (quotient) and remainder = 7.

53 ÷ 9 =
5 (quotient) and remainder = 8 (5 + 3).

75 ÷ 9 =
7 (quotient) and remainder = 12 (7 + 5).
But 12 is more than 9.
9 goes 'once' into 12 and leaves a remainder of 3.
So we add '1' to the earlier quotient (7) and that gives us the answer.

75 ÷ 9 = 8 (quotient) and remainder = 3.

98 ÷ 9 =
9 (quotient) and remainder = 17.
17 is more than 9.
9 goes 'once' into 17 and leaves a remainder of 8.
So we add '1' to the earlier quotient (9) and that gives us the answer.

98 ÷ 9 = 10 (quotient) and remainder = 8.

Sidebar

In mathematical terms, the number being divided is called the DIVIDEND, the number that is dividing is called the DIVISOR. QUOTIENT is the number of times the divisor can divide into the dividend and what remains is called the REMAINDER.

So DIVIDEND = DIVISOR X QUOTIENT + REMAINDER.

> **Rule**
> Now lets divide three digit numbers by 9
>
> 1. In the quotient line, write down the left most digit and then the sum of the left two digits. If the sum of the digits is more than 9, keep the right side digit and shift the left side digit to the left.
> 2. Sum up all the digits and that is the remainder. If this is more than 9, <u>see how many times 9 divides into this and shift</u> that to the left and keep the net remainder as the remainder.

Illustration

121 ÷ 9 =
Quotient = 13 (we put down 1 and then the sum of the left side digits: 1 + 2.)
Remainder = 1 + 2 + 1 = 4.

121 ÷ 9 = 13 (quotient) and remainder = 4.

251 ÷ 9 =
Quotient = 27 (2 and 2 + 5)
Remainder = 8 (Sum of the digits = 2 + 5 + 1)

251 ÷ 9 = 27 (quotient) and remainder = 8.

563 ÷ 9 =
Quotient = 5_11 (5 + 6 = 11 and is double digit. Keep the right side digit and shift the left side digit ie. 5_11 = 61.)
Quotient = 61
Remainder = 5 + 6 + 3 = 14 (This is more than 9 and 9 goes 'once' into 14 and gives a remainder of 5. So we will keep 5 and shift '1' to the quotient.)
That makes the quotient = 61 + 1 = 62 and remainder = 5.

563 ÷ 9 = 62 (quotient) and remainder = 5.

842 ÷ 9 =
Quotient = 8_12 (8 + 4 = 12. Shift 1 keep 2.)
Quotient = 92
Remainder = 8 + 4 + 2 = 14 (14 is greater than 9. Shift '1' to the quotient and keep 5.)
Quotient = 92 + 1 = 93

842 ÷ 9 = 93 (quotient) and remainder = 5.

Rule
Dividing larger numbers by 9.

We will just follow the same pattern.

1. From the left, put down the first digit and then sum of digits into the quotient line till you reach the last digit. Any time the sum is a double digit, keep the right side digit and shift the left side digit to the left.
2. Sum up all the digits and that is the remainder. If this is more than 9, see how many times 9 divides into this and shift that to the quotient and keep the net remainder as the remainder.

Illustration

231121 ÷ 9 =
Lets do sum of digits starting from the left

2 =	2	(first digit)
2 + 3 =	5	(sum of first 2 digits.)
2 + 3 + 1 =	6	(sum of first 3 digits.)
2 + 3 + 1 + 1 =	7	(sum of first 4 digits.)
2 + 3 + 1 + 1 + 2 =	9	(sum of first 5 digits. This is till the last digit.)

So Quotient = 25679

2 + 3 + 1 + 1 + 2 + 1 =   10   (Sum of all the digits is the remainder. This is more than 9. And 9 goes 'once' into 10 and leaves a remainder of 1.)

So we will shift this '1' to the quotient and keep 1 as remainder.

Quotient = 25679 + 1 = 25680.

231121 ÷ 9 = 25680 (quotient) and remainder = 1.

Lets do one example with lesser explanation:
1213123 ÷ 9 =

Quotient: 1 3 4 7 8 $_1$0 = 134790.

Remainder = 1 + 2 + 1 + 3 + 1 + 2 + 3 = 13. 9 goes 'once' into 13 and leaves remainder 4.

Quotient = 134790 + 1 = 134791

1213123 ÷ 9 = 134791 (quotient) and remainder = 4.

Exercise:

Write the Quotient and the Remainder in the boxes:

	Quotient	Remainder
24 ÷ 9		
44 ÷ 9		
71 ÷ 9		
80 ÷ 9		
115 ÷ 9		
129 ÷ 9		
346 ÷ 9		
484 ÷ 9		
1441 ÷ 9		
2341 ÷ 9		
6441 ÷ 9		
8765 ÷ 9		
11111 ÷ 9		

	Quotient	Remainder
12345 ÷ 9	1371	6
136523 ÷ 9	15169	2
982344 ÷ 9	109149	3
1123122 ÷ 9	124791	3

Shortcut 34: Multiplying numbers close to and below 100

Rule
1. Calculate the difference of both the numbers from 100.
2. Subtract the difference from 100 of <u>any one number from the other number</u>. This is the left digits of the answer.
3. Multiply the difference from 100 of both the numbers. This will be the <u>right two digits</u> of the answer. Put these after the result of step 2 to get the answer.

Illustration
96 X 93
1. Difference of the numbers from 100 are: 4 (100 – 96) and 7 (100 – 93)
2. We subtract 4 from 93 or 7 from 96 (answer will be same).
   93 – 4 = 89 or 96 – 7 = 89
   89 is the left two digits of the answer.
3. Multiply 7 X 4 and add it to the result of step 2.
   7 X 4 = 28
   = 8928

96 X 93 = 8928

Examples:
97 X 94
1. 100 – 97 = 3 and 100 – 94 = 6
2. 97 – 6 = 91 (subtract the difference from 100 of one number from the other number. Also note that 94 – 3 = 91.)
   91 is the left two digits of the answer.
3. 3 X 6 = 18 (multiply the difference from 100 of both the numbers. This forms the right two digits of the answer.)
   So 97 X 94 = 9118

99 X 98
1. 100 – 99 = 1 and 100 – 98 = 2
2. 99 – 2 = 97 (subtract the difference from 100 of one number from the other number. Also note that 98 – 1 = 97.)
   97 is the left two digits of the answer.
3. 1 X 2 = 2 [This should form the right two digits of the answer. But this is a single digit result. So we will put a '0' (zero) before it to make it 02].

   So 97 X 94 = 9702

Examples:

92 X 93
1. 100 − 92 = 8 and 100 − 93 = 7.
2. 92 − 7 = 85 (subtract the difference from 100 of one number from the other number. Also note that 93 - 8 = 85.)
   85 is the left two digits of the answer
3. 7 X 8 = 56 (multiply the difference from 100 of both the numbers. This forms the right two digits of the answer.)
   So 92 X 93 = 8556

86 X 84
1. 100 − 86 = 14 and 100 − 84 = 16.
2. 86 − 16 = 70 (subtract the difference from 100 of one number from the other number.)
   70 is the left two digits of the answer.
3. 16 X 14 = 224 [This should form the right two digits of the answer. But these are three-digits. So we will take the right two digits and carry over the remaining digit (2) and add that to 70.]
   $70_2 24 = 7224$

   So 86 X 84 = 7224

96 X 94
1. 100 − 96 = 4 and 100 − 94 = 6
2. 96 − 6 = 90 Or 94 − 4 = 90
3. 6 X 4 = 24

So 96 X 94 = 9024

97 X 97
1. 100 − 97 = 3 and 100 − 97 = 3
2. 97 − 3 = 94
3. 3 X 3 = 9 (so we make it 09.)

So 97 X 97 = 9409

88 X 90
1. 100 − 88 = 12 and 100 − 90 = 10
2. 88 − 10 = 78
3. 12 X 10 = 120 (we keep the 20 and carry the digit '1' to the left side digits.)

$78_1 20 = 7920$

So 88 X 90 = 7920

Exercise:

1) 95 X 92 =

Difference from 100

100 - 95 = [ ]

100 - 92 = [ ]

Subtract difference of any number from the other number.

[ ] - [ ] = [ ]

Multiply the difference from 100 of both the numbers.

[ ] X [ ] = [ ]

Put the difference result as the left two digits and the multiplication result as the right two digits.

95 X 92 = [ ]

---

2) 94 X 95 =

Difference from 100

100 - 94 = [ ]

100 - 95 = [ ]

Subtract difference of any number from the other number.

[ ] - [ ] = [ ]

Multiply the difference from 100 of both the numbers.

[ ] X [ ] = [ ]

Put the difference result as the left two digits and the multiplication result as the right two digits.

94 X 95 = [ ]

3) 92   X   91   =

Difference from 100

100 − 92 = ☐

100 − 91 = ☐

Subtract difference of any number from the other number.

☐ − ☐ = ☐

Multiply the difference from 100 of both the numbers.

☐ X ☐ = ☐

Put the difference result as the left two digits and the multiplication result as the right two digits.

92   X   91   =   ☐

---

4) 98   X   97   =

Difference from 100

100 − 98 = ☐

100 − 97 = ☐

Subtract difference of any number from the other number.

☐ − ☐ = ☐

Multiply the difference from 100 of both the numbers.

☐ X ☐ = ☐

Note: This has to form the <u>right two digits.</u>
Since this is a single digit result, remember to put a 'zero' before it.

Put the difference result as the left two digits and the multiplication result as the right two digits.

98   X   97   =   ☐

Exercises: Solve the following questions in the table below. The first one is done as an example:

	Difference from 100 of both the numbers		Subtract difference from 100 of one number from the other number	Product of the differences	Answer
94 X 92	6	8	86	48	8648
	(100−94)	(100−92)	(94−8)	(6 X 8)	(86 and 48)
96 X 91					
98 X 95					
97 X 93					
96 X 92					
98 X 92					
99 X 91					
99 X 96					
98 X 98					
92 X 85					
95 X 80					

Shortcut 35: Multiplying numbers close to and above 100

This is very similar to shortcut # 34.

Rule
1. Calculate the difference of both the numbers from 100.
2. *Add* the difference from 100 of <u>any one number to the other number.</u> This is the left three digits of the answer.
3. Multiply the difference from 100 of both the numbers. This will be the right two digits of the answer. Put these after the result of step 2 to get the answer.

Illustration

106 X 103
1. Difference of the numbers from 100 are: 6 (106 – 100) and 3 (103 – 100).
2. We add 6 to 103 or 3 to 106 (answer will be same).
    106 + 3 = 109 or 103 + 6 = 109
    109 is the left three digits of the answer.
3. Multiply 6 X 3 and add it to the result of step 2.
    6 X 3 = 18
    = 10918

106 X 103 = 10918

Examples:
107 X 104
1. 107 – 100 = 7 and 104 - 100 = 4
2. 107 + 4 = 111 (Add the difference from 100 of one number from the other number. Also note that 104 + 7 = 111.)
    111 is the left three digits of the answer.
3. 7 X 4 = 28 (multiply the difference from 100 of both the numbers. This forms the right two digits of the answer.)
So 107 X 104 = 11128.

108 X 107
1. 108 – 100 = 8 and 107 – 100 = 7
2. 108 + 7 = 115 (also 107 + 8 = 115)
3. 7 X 8 = 56 (multiply the difference from 100 of both the numbers. This forms the right two digits of the answer.)

So 108 X 107 = 11556.

Examples:

102 X 101
1. 102 − 100 = 2 and 101 − 100 = 1
2. 102 + 1 = 103
3. 1 X 2 = 2 [This should form the right two digits of the answer. But this is a single digit result. So we will put a '0' (zero) before it to make it 02.]

So 102 X 101 = 10302.

103 X 103
1. 103 − 100 = 3 and 103 − 100 = 3
2. 103 + 3 = 106
3. 3 X 3 = 9 (so we make it 09.)

So 103 X 103 = 10609.

106 X 104
1. 106 − 100 = 6 and 104 − 100 = 4
2. 106 + 4 = 110 Or 104 + 6 = 110
3. 6 X 4 = 24

So 106 X 104 = 11024.

112 X 110
1. 112 − 100 = 12 and 110 − 100 = 10
2. 112 + 10 = 122
3. 12 X 10 = 120, we keep the 20 and carry the '1' to the left side digits.
   122$_1$20 = 12320

So 112 X 110 = 12320.

116 X 114
1. 116 − 100 = 16 and 114 − 100 = 14
2. 116 + 14 = 130
3. 16 X 14 = 224 [This should form the right two digits of the answer. But these are three-digits. So we will take the right two digits and carry over the remaining digit (2) and add that to 130.]
   130$_2$24 = 13224

So 116 X 114 = 13224.

Exercise:

1) 105 X 108 =

Difference from 100

105 - 100 = [ ]

108 - 100 = [ ]

Add difference of any number to the other number.

[ ] + [ ] = [ ]

Multiply the difference from 100 of both the numbers.

[ ] X [ ] = [ ]

Put the difference result as the left two digits and the multiplication result as the right two digits.

105 X 108 = [ ]

---

2) 106 X 105 =

Difference from 100

106 - 100 = [ ]

105 - 100 = [ ]

Add difference of any number to the other number.

[ ] + [ ] = [ ]

Multiply the difference from 100 of both the numbers.

[ ] X [ ] = [ ]

Put the difference result as the left two digits and the multiplication result as the right two digits.

106 X 105 = [ ]

3) 108   X   109   =

Difference from 100

108 − 100 = ☐

109 − 100 = ☐

Add difference of any number to the other number.

☐ + ☐ = ☐

Multiply the difference from 100 of both the numbers.

☐ X ☐ = ☐

Put the difference result as the left digits and the multiplication result as the right two digits.

92 X 91 = ☐

---

4) 102   X   103   =

Difference from 100

102 − 100 = ☐

103 − 100 = ☐

Add difference of any number to the other number.

☐ + ☐ = ☐

Multiply the difference from 100 of both the numbers.

☐ X ☐ = ☐

Note: This has to form the <u>right two digits.</u> Since this is a single digit result, remember to put a 'zero' before it.

Put the difference result as the left digits and the multiplication result as the right two digits.

102 X 103 = ☐

5) 110   X   115   =

Difference from 100

110 - 100 = ☐

115 - 100 = ☐

Add difference of any number to the other number.

☐ + ☐ = ☐

Multiply the difference from 100 of both the numbers.

☐ X ☐ = ☐

Note: This has to form the right two digits. Since this is a single digit result, remember to put a 'zero' before it.

Put the difference result as the left digits and the multiplication result as the right two digits.

110   X   115   = ☐

Exercises: Solve the following questions in the table below. The first one is done as an example:

	Difference from 100 of both the numbers		Add difference from 100 of one number to the other number	Product of the differences	Answer
106 X 108	6	8	114	48	11448
	(106 − 100)	(108 −100)	(106 + 8)	(6 X 8)	(114 and 48)
104 X 109	4	9	113	36	11336
102 X 105	2	5	107	10	10710
103 X 107	3	7	110	21	11021
104 X 108	4	8	112	32	11232
102 X 108	2	8	110	16	11016
101 X 109	1	9	110	9	11009
101 X 104	1	4	105	4	10504
102 X 102	2	2	104	4	10404
108 X 115	8	15	123	120	12420
105 X 120	5	20	125	100	12600

Shortcut 36: Multiplying numbers close to and below 1000

Rule
1. Calculate the difference of both the numbers from 1000.
2. Subtract the difference from 1000 of <u>any one number from the other number.</u> This is the left digits of the answer.
3. Multiply the difference from 1000 of both the numbers. This will be the <u>right three digits</u> of the answer. Put these after the result of step 2 to get the answer.

Important thing to note here is that since we are multiplying numbers near 1000, the result in step 3 should form the <u>three-right-digits</u> of the answer. If it is one-digit or two-digit result, we should add leading 'zeroes' to make it three-digit result. If however, it is bigger than three digits, we need to carry the digit other than the three right digits.

Illustration

985 X 990
1. Difference of the numbers from 1000 are: 15 (1000 – 985) and 10 (1000 – 990).
2. We subtract 15 from 990 or 10 from 985 (answer will be same).
   985 – 10 = 975 or 990 – 15 = 975
   975 is the left side digits of the answer.
3. Multiply 15 X 10 and add it to the result of step 2.
   15 X 10 = 150
   = 975150

985 X 990 = 975150

Examples:
987 X 989
1. 1000 – 987 = 13 and 1000 – 989 = 11
2. 987 – 11 = 976 (subtract the difference from 1000 of one number from the other number. Also note that 989 – 13 = 976.)
   976 is the left side digits of the answer.
3. 13 X 11 = 143 (multiply the difference from 1000 of both the numbers. This forms the right three digits of the answer.)

So 987 X 989 = 976143.

992 X 980
1. 1000 – 992 = 8 and 1000 – 980 = 20
2. 992 – 20 = 972 (subtract the difference from 1000 of one number from the other number.)
   972 is the left digits of the answer.
3. 8 X 20 = 160 (multiply the difference from 100 of both the numbers. This forms the right three digits of the answer.)

So 992 X 980 = 972160.

Examples:
995 X 985
1. 1000 − 995 = 5 and 1000 − 985 = 15
2. 995 − 15 = 980 (subtract the difference from 1000 of one number from the other number. Also note that 985 − 5 = 980.)
   980 is the left side digits of the answer.
3. 15 X 5 = 75 [This should form the right three digits of the answer. But this is a two-digit result. So we will put a '0' (zero) before it to make it 075.]

So 995 X 985 = 980075.

986 X 984
1. 1000 − 986 = 14 and 1000 − 984 = 16
2. 986 − 16 = 970
   970 is the left three digits of the answer.
3. 16 X 14 = 224 (These are the right three digits of the answer.)

So 986 X 984 = 970224.

996 X 994
1. 1000 − 996 = 4 and 1000 − 994 = 6
2. 996 − 6 = 990 Or 994 − 6 = 990
3. 6 X 4 = 24 (This is two digits so we will put a leading 'zero' and make it 024.)

So 996 X 994 = 990024.

997 X 997
1. 1000 − 997 = 3 and 1000 − 997 = 3
2. 997 − 3 = 994
3. 3 X 3 = 9 (so we make it 009.)

So 97 X 97 = 994009.

950 X 925
1. 1000 − 950 = 50 and 1000 − 925 = 75
2. 950 − 75 = 875
3. 50 X 75 = 3750 (Since this is a four digit result, we will keep the 750 and carry the digit '3' to the left side digits.)
875₃750 = 878750

So 950 X 925 = 878750.

Exercise:

1) 995   X   992   =

Difference from 1000

1000 − 995 = [   ]

1000 − 992 = [   ]

Subtract difference of any number from the other number.

[   ] − [   ] = [   ]

Multiply the difference from 1000 of both the numbers.

[   ] X [   ] = [   ]

Put the difference result as the left side digits and the multiplication result as the right side three digits.

995 X 992 = [   ]

---

2) 994   X   995   =

Difference from 1000

1000 − 994 = [   ]

1000 − 995 = [   ]

Subtract difference of any number from the other number.

[   ] − [   ] = [   ]

Multiply the difference from 1000 of both the numbers.

[   ] X [   ] = [   ]   Put leading 'zeroes' to make it a three digit result.

Put the difference result as the left side digits and the multiplication result as the right side three digits.

994 X 995 = [   ]

3) 992   X   991   =

Difference from 1000

1000 − 992 = ☐

1000 − 991 = ☐

Subtract difference of any number from the other number.

☐ − ☐ = ☐

Multiply the difference from 1000 of both the numbers.

☐ X ☐ = ☐

Put the difference result as the left side digits and the multiplication result as the right side three digits.

992 X 991 = ☐

---

4) 998   X   997   =

Difference from 1000

1000 − 998 = ☐

1000 − 997 = ☐

Subtract difference of any number from the other number.

☐ − ☐ = ☐

Multiply the difference from 1000 of both the numbers.

☐ X ☐ = ☐

Note: Remember to put leading 'zeroes' to make it a three digit result.

Put the difference result as the left side digits and the multiplication result as the right side three digits.

998 X 997 = ☐

5) 990   X   975   =

Difference from 1000

1000 − 990 = [10]

1000 − 975 = [25]

Subtract difference of any number from the other number.

[990] − [25] = [965]

Multiply the difference from 1000 of both the numbers.

[10] X [25] = [250]

Put the difference result as the left side digits and the multiplication result as the right side three digits.

990 X 975 = [965250]

Exercises: Solve the following questions in the table below. The first one is done as an example:

	Difference from 1000 of both the numbers		Subtract difference from 1000 of one number from the other number	Product of the differences	Answer
994 X 992	6	8	986	48	986048
	(1000 − 994)	(1000 − 992)	(994 − 8)	(6 X 8)	(986 and 048)
996 X 991	4	9	987	36	987036
998 X 995	2	5	993	10	993010
997 X 993	3	7	990	21	990021
996 X 982	4	18	978	72	978072
989 X 992	11	8	981	88	981088
999 X 991	1	9	990	9	990009
998 X 998	2	2	996	4	996004
981 X 995	19	5	976	95	976095
992 X 985	8	15	977	120	977120
955 X 980	45	20	935	900	935900

Shortcut 37: Multiplying numbers close to and above 1000

Rule
1. Calculate the difference of both the numbers from 1000.
2. Add the difference from 1000 of <u>any one number to the other number.</u> This is the left digits of the answer.
3. Multiply the difference from 1000 of both the numbers. This will be the <u>right three digits</u> of the answer. Put these after the result of step 2 to get the answer.

Important thing to note here is that since we are multiplying numbers near 1000, the result in step 3 should form the three-right-digits of the answer. If it is one digit or two digit result, we should add leading 'zeroes' to make it three digit result. If however, it is bigger than three digits, we need to carry the digit other than the three digits to the left side.

Illustration

1015 X 1010
1. Difference of the numbers from 1000 are: 15 (1015 – 1000) and 10 (1010 – 1000).
2. We add 15 to 1010 or 10 to 1015 (answer will be same).
   1010 + 15 = 1025 or 1015 + 10 = 1025
   1025 is the left side digits of the answer.
3. Multiply 15 X 10 and add it to the result of step 2.
   15 X 10 = 150
   = 1025150

1015 X 1010 = 1025150

Examples:
1013 X 1011
1. 1013 – 1000 = 13 and 1011 – 1000 = 11
2. 1013 + 11 = 1024 (add the difference from 1000 of one number to the other number. Also note that 1011 + 13 = 1024.)
   1024 is the left side digits of the answer.
3. 13 X 11 = 143 (multiply the difference from 1000 of both the numbers. This forms the right three digits of the answer.)

So 1013 X 1011 = 1024143.

1008 X 1020
1. 1008 – 1000 = 8 and 1020 – 1000 = 20
2. 1008 + 20 = 1028 (add the difference from 1000 of one number to the other number. Also note that 1020 + 8 = 1028.)
   1028 is the left digits of the answer.
3. 8 X 20 = 160 (multiply the difference from 1000 of both the numbers. This forms the right three digits of the answer.)

So 1008 X 1020 = 1028160.

Examples:
1005 X 1015
1. 1005 − 1000 = 5 and 1015 − 1000 = 15
2. 1005 + 15 = 1020 (add the difference from 1000 of one number to the other number. Also note that 1015 + 5 = 1020.)
    1020 is the left side digits of the answer.
3. 15 X 5 = 75 [This should form the right three digits of the answer. But this is a two-digit result. So we will put a leading '0' (zero) before it to make it 075.]

So 1005 X 1015 = 1020075.

1014 X 1016
1. 1014 − 1000 = 14 and 1016 − 1000 = 16
2. 1014 + 16 = 1030
    1030 is the left digits of the answer.
3. 16 X 14 = 224 (These are the right three digits of the answer.)

So 1014 X 1016 = 1030224.

1006 X 1004
1. 1006 − 1000 = 6 and 1004 − 1000 = 4
2. 1006 + 4 = 1010 Or 1004 + 6 = 1010
3. 6 X 4 = 24 (This is two digits so we will put a leading 'zero' and make it 024.)

So 1006 X 1004 = 1010024.

1003 X 1003
1. 1003 − 1000 = 3 and 1003 − 1000 = 3
2. 1003 + 3 = 1006
3. 3 X 3 = 9 (so we make it 009.)

So 1003 X 1003 = 1006009.

1050 X 1075
1. 1050 − 1000 = 50 and 1075 − 1000 = 75
2. 1050 + 75 = 1125
3. 50 X 75 = 3750 (Since this is a four digit result, we will keep the 750 and carry the '3' to the left side digits.)
    1125₃750 = 1128750

So 1050 X 1025 = 1128750.

Exercise:

1) 1005   X   1008   =

Difference from 1000

1005 − 1000 = ☐

1008 − 1000 = ☐

Add difference of any number to the other number.

☐ + ☐ = ☐

Multiply the difference from 1000 of both the numbers.

☐ X ☐ = ☐   Put a leading 'zero' to make it a three digit result.

Put the difference result as the left side digits and the multiplication result as the right side three digits.

1005   X   1008   = ☐

---

2) 1006   X   1005   =

Difference from 1000

1006 − 1000 = ☐

1005 − 1000 = ☐

Add difference of any number to the other number.

☐ + ☐ = ☐

Multiply the difference from 1000 of both the numbers.

☐ X ☐ = ☐   Put a leading 'zero' to make it a three digit result.

Put the difference result as the left side digits and the multiplication result as the right side three digits.

1006   X   1005   = ☐

3) 1008   X   1009   =

Difference from 1000

1008 − 1000 = ☐

1009 − 1000 = ☐

Subtract difference of any number to the other number.

☐ + ☐ = ☐

Multiply the difference from 1000 of both the numbers.

☐ X ☐ = ☐   Put a leading 'zero' to make it a three digit result.

Put the difference result as the left side digits and the multiplication result as the right side three digits.

1008   X   1009   = ☐

---

4) 1002   X   1003   =

Difference from 1000

1002 − 1000 = ☐

1003 − 1000 = ☐

Add difference of any number to the other number.

☐ + ☐ = ☐

Multiply the difference from 1000 of both the numbers.

☐ X ☐ = ☐

Note: Remember to put leading 'zeroes' to make it a three digit result.

Put the difference result as the left two digits and the multiplication result as the right side three digits.

1002   X   1003   = ☐

5) 1004 X 1011 =

Difference from 1000

1004 − 1000 = [ ]

1011 − 1000 = [ ]

Add difference of any number to the other number.

[ ] + [ ] = [ ]

Multiply the difference from 1000 of both the numbers.

[ ] X [ ] = [ ]   Put a leading 'zero' to make it a three digit result.

Put the difference result as the left side digits and the multiplication result as the right side three digits.

1004 X 1011 = [ ]

---

6) 1010 X 1075 =

Difference from 1000

1010 − 1000 = [ ]

1075 − 1000 = [ ]

Add difference of any number to the other number.

[ ] + [ ] = [ ]

Multiply the difference from 1000 of both the numbers.

[ ] X [ ] = [ ]

Note: Remember to put leading 'zeroes' to make it a three digit result.

Put the difference result as the left side digits and the multiplication result as the right side three digits.

1010 X 1075 = [ ]

Exercises: Solve the following questions in the table below. The first one is done as an example:

	Difference from 1000 of both the numbers		Add difference from 1000 of one number to the other number	Product of the differences	Answer
1006 X 1008	6	8	1014	48	1014048
	(1006−1000)	(1008−1000)	(1006 + 8)	(6 X 8)	(1014 and 048)
1004 X 1009					
1002 X 1005					
1003 X 1007					
1004 X 1018					
1011 X 1008					
1001 X 1009					
1002 X 1002					
1019 X 1005					
1008 X 1015					
1045 X 1020					

Shortcut 38: Squaring numbers near 100

Rule
Calculate the difference from 100 of the number to be squared.

If the number to be squared is less than 100, subtract the difference from the number being squared.
If the number to be squared is more than 100, add the difference to the number being squared.
This will form the left side digits of the answer.

Square the difference from 100 and put that as the two right side digits of the answer. As before, if the square is a single digit answer, put a '0' (zero) before it and if it is a three digit answer, take the right two digits and carry the third digit to the left side.

Illustration

Calculate 94^2
$100 - 94 = 6$
Since 94 is less than 100, subtract 6 from 94:
$94 - 6 = 88$ (left two digits of the answer.)

$6^2 = 36$ (right two digits of the answer.)

$94^2 = 8836$

Examples:
Calculate 99^2
$100 - 99 = 1$
Since 99 is less than 100, subtract 1 from 99:
$99 - 1 = 98$

$1^2 = 1$ (since it is single digit, we put a leading 'zero' before the 1.)

$99^2 = 9801$

Calculate 88^2
$100 - 88 = 12$
Since 88 is less than 100, subtract 12 from 88:
$88 - 12 = 76$

$12^2 = 144$ (since it is three digits so we carry the digit '1' to the left side.)

$88^2 = 76_144$
$88^2 = 7744$

Calculate 104²
104 - 100 = 4
Since 104 is more than 100, add 4 to 104:
104 + 4 = 108

4² = 16

104² = 10816

Calculate 101²
101 - 100 = 1
Since 101 is more than 100, add 1 to 101:
101 + 1 = 102

1² = 1 (we put a leading 'zero' before the 1.)

101² = 10201

Calculate 112²
112 - 100 = 12
Since 112 is more than 100, add 12 to 112:
112 + 12 = 124

12² = 144 (we carry the digit '1' to the left side.)

112² = 124₁44
112² = 12544

Exercise:

1) $92^2 =$

a) $92 - \boxed{} = \boxed{}$
      $(100 - 92)$

b) $\boxed{} \times \boxed{} = \boxed{}$
  $(100 - 92)$ $(100 - 92)$

$92^2 = \boxed{}$
Put the result of a) and then the result of b).

---

2) $96^2 =$

a) $96 - \boxed{} = \boxed{}$
      $(100 - 96)$

b) $\boxed{} \times \boxed{} = \boxed{}$
  $(100 - 96)$ $(100 - 96)$

$96^2 = \boxed{}$

---

3) $98^2 =$

a) $98 - \boxed{} = \boxed{}$

b) $\boxed{} \times \boxed{} = \boxed{}$

$98^2 = \boxed{}$ (remember to put a leading 'zero' in the result of step b.)

4) 87^2 =

a) 87 − ☐ = ☐

b) ☐ × ☐ = ☐

87^2 = ☐

---

5) 108^2 =

a) 108 + ☐ = ☐
              (108 − 100)

b) ☐ × ☐ = ☐
   (108 − 100)   (108 − 100)

108^2 = ☐
Put the result of a) and then the result of b).

---

6) 106^2 =

a) 106 + ☐ = ☐
              (106 − 100)

b) ☐ × ☐ = ☐
   (106 − 100)   (106 − 100)

106^2 = ☐

7) $102^2 =$

a) 102 + [ ] = [ ]

b) [ ] × [ ] = [ ]

102^2 = [ ] (remember to put a leading 'zero' in the result of step b.)

---

8) $113^2 =$

a) 113 + [ ] = [ ]

b) [ ] × [ ] = [ ]

113 = [ ] (remember to carry the digits.)

Exercises:

For the following, write down the answer directly:

1) 97^2 = ☐

2) 95^2 = ☐

3) 93^2 = ☐

4) 89^2 = ☐

5) 103^2 = ☐

6) 84^2 = ☐

7) 91^2 = ☐

8) 107^2 = ☐

9) 115^2 = ☐

10) 86^2 = ☐

11) 109^2 = ☐

12) 89^2 = ☐

13) 111^2 = ☐

Shortcut 39: Squaring numbers near 1000

Rule
Calculate the difference from 1000 of the number to be squared.

If the number to be squared is less than 1000, subtract the difference from the number being squared.
If the number to be squared is more than 1000, add the difference to the number being squared.
This will form the left side digits of the answer.

Square the difference from 1000 and put that as the three right side digits of the answer. As before, if the square is a single digit answer, put '00' (two leading zeroes) before it, if it is a two digit number, put one leading '0' (zero) before it and if it is a four digit answer, take the right three digits and carry the extra digit to the left side.

Illustration

Calculate 988^2
$1000 - 988 = 12$
Since 988 is less than 1000, subtract 12 from 988:
$988 - 12 = 976$

$12^2 = 144$

$988^2 = 976144$

Calculate 999^2
$1000 - 999 = 1$
Since 999 is less than 1000, subtract 1 from 999:
$999 - 1 = 998$

$1^2 = 1$ (we put two leading 'zeroes' before the 1.)

$999^2 = 998001$

Calculate 925^2
$1000 - 925 = 75$
Since 925 is less than 1000, subtract 75 from 925:
$925 - 75 = 850$

$75^2 = 5625$ (we carry the digit '5' to the left side since we can keep only three digits.)

$925^2 = 850_5 625$

$925^2 = 855625$

Illustration

Calculate 898²
1000 − 898 = 102
Since 898 is less than 1000, subtract 102 from 898:
898 − 102 = 796

102² = 10404 (we carry the digit 10 to the left side since we can keep only three digits.)

898² = 796$_{10}$404
898² = 806404

Calculate 1012²
1012 − 1000 = 12
Since 1012 is more than 1000, add 12 to 1012.
1012 + 12 = 1024

12² = 144

1012² = 1024144

Calculate 1001²
1001 − 1000 = 1
Since 1001 is more than 1000, add 1 to 1001:
1001 + 1 = 1002

1² = 1 (we put two leading 'zeroes' before the 1.)

1001² = 1002001

Calculate 1075²
1075 − 1000 = 75
Since 1075 is more than 1000 add 75 to 1075:
1075 + 75 = 1150

75² = 5625 (we carry the digit 5 to the left side since we can keep only three digits.)

1075² = 1150$_5$625

1075² = 1155625

Illustration

Calculate 1112²

1112 − 1000 = 112
Since 1112 is more than 1000 add 112 to 1112:
1112 + 112 = 1224

112² = 12544 (we carry the '12' to the left side since we can keep only three digits.)

1112² = 1224₁₂544

1112² = 1236544

Exercise:

1) $992^2 =$

a) $992 - \boxed{} = \boxed{}$
         $(1000 - 992)$

b) $\boxed{} \times \boxed{} = \boxed{}$
   $(1000 - 992)$ $(1000 - 992)$

$992^2 = \boxed{}$
Put the result of a) and then the result of b).
(remember to put leading 'zeroes' in the result of step b.)

---

2) $986^2 =$

a) $986 - \boxed{} = \boxed{}$
         $(1000 - 986)$

b) $\boxed{} \times \boxed{} = \boxed{}$
   $(1000 - 986)$ $(100 - 986)$

$986^2 = \boxed{}$

---

3) $998^2 =$

a) $998 - \boxed{} = \boxed{}$

b) $\boxed{} \times \boxed{} = \boxed{}$

$998^2 = \boxed{}$ (remember to put a leading 'zero' in the result of step b.)

4) $1005^2 =$

a) $1005 + \boxed{}\ =\ \boxed{}$
(1005 − 1000)

b) $\boxed{} \times \boxed{}\ =\ \boxed{}$
(1005 − 1000) (1005 − 1000)

$1005^2\ =\ \boxed{}$
Put the result of a) and then the result of b).
(remember to put leading 'zeroes' in the result of step b.)

---

5) $1014^2 =$

a) $1014 + \boxed{}\ =\ \boxed{}$
(1014 − 1000)

b) $\boxed{} \times \boxed{}\ =\ \boxed{}$
(1014 − 1000) (1014 − 1000)

$1014^2\ =\ \boxed{}$

---

6) $1050^2 =$

a) $1050 + \boxed{}\ =\ \boxed{}$

b) $\boxed{} \times \boxed{}\ =\ \boxed{}$

$1050^2\ =\ \boxed{}$ (remember to keep only three digits from the result of step b) and carry the rest to the left side.)

7) $1008^2 =$

a) $1008 + \boxed{} = \boxed{}$
   $(1008 - 1000)$

b) $\boxed{} \times \boxed{} = \boxed{}$
   $(1008 - 1000)\ (1008 - 1000)$

$1008^2 = \boxed{}$
Put the result of a) and then the result of b).

---

8) $1006^2 =$

a) $1006 + \boxed{} = \boxed{}$
   $(1006 - 1000)$

b) $\boxed{} \times \boxed{} = \boxed{}$
   $(1006 - 1000)\ (1006 - 1000)$

$1006^2 = \boxed{}$

---

9) $1002^2 =$

a) $1002 + \boxed{} = \boxed{}$

b) $\boxed{} \times \boxed{} = \boxed{}$

$1002^2 = \boxed{}$ (remember to put leading 'zeroes' in the result of step b to make it a three-digits result.)

10) $1113^2 =$

a) $1113 + $ ☐ $=$ ☐

b) ☐ $\times$ ☐ $=$ ☐

$1113^2 \quad = $ ☐
(remember to carry the digit.)

Exercises:

For the following, write down the answer directly:

1) 997^2 = ☐

2) 995^2 = ☐

3) 993^2 = ☐

4) 984^2 = ☐

5) 1003^2 = ☐

6) 991^2 = ☐

7) 992^2 = ☐

8) 1007^2 = ☐

9) 1015^2 = ☐

10) 1018^2 = ☐

11) 1009^2 = ☐

12) 1011^2 = ☐

13) 1111^2 = ☐

Shortcut 40: Squaring numbers near 200

Rule
*This is very similar to squaring numbers near 100.
Only added step is that we double the left side digits.*

Calculate the difference from 200 of the number to be squared.

*If the number to be squared is less than 200, subtract the difference from the number being squared.
If the number to be squared is more than 200, add the difference to the number being squared.*

Now, multiply this answer by 2 - Added step.

This will form the left side digits of the answer.

Square the difference from 200 and put that as the two right side digits of the answer. As before, if the square is a single digit answer, put a leading '0' (zero) before it and if it is a three digit answer, take the right two digits and carry the third digit to the left side.

Illustration

Calculate 194^2
200 − 194 = 6
Since 194 is less than 200, subtract 6 from 194:
194 − 6 = 188

188 X 2 = 376 (this is the added step compared to squaring numbers near 100.)

6^2 = 36

194^2 = 37636

Calculate 199^2
200 − 199 = 1
Since 199 is less than 200, subtract 1 from 199:
199 − 1 = 198

198 X 2 = 396

1^2 = 1 (since it is single digit, we put a leading 'zero' before the 1.)

199^2 = 39601

Illustration

Calculate 188²
200 − 188 = 12
Since 188 is less than 200, subtract 12 from 188:
188 − 12 = 176

176 X 2 = 352

12^2 = 144 (we carry '1' to the left side.)

188^2 = 352₁44

188^2 = 35344

Calculate 204²
204 − 200 = 4
Since 204 is more than 200, add 4 to 204:
204 + 4 = 208

208 X 2 = 416

4^2 = 16

204^2 = 41616

Calculate 201²
201 − 200 = 1

201 + 1 = 202

202 x 2 = 404

1^2 = 1 (we put a leading 'zero' before the 1.)

201^2 = 40401

Calculate 212²
212 − 200 = 12

212 + 12 = 224

224 X 2 = 448

12^2 = 124 (we carry the digit '1' to the left side.)

212^2 = 448₁44

212^2 = 44944

Exercise:

1) 192^2 =

a)   192 − ☐   =   ☐   X 2 = ☐
            (200 − 192)   double

b)   ☐   X   ☐   =   ☐
     (200 − 192)   (200 − 192)

     192^2 = ☐
           Put the result of a) and then the result of b).

---

2) 196^2 =

a)   196 − ☐   =   ☐   X 2 = ☐
            (200 − 196)   double

b)   ☐   X   ☐   =   ☐
     (200 − 196)   (200 − 196)

     196^2 = ☐

---

3) 198^2 =

a)   198 − ☐   =   ☐   X 2 = ☐

b)   ☐   X   ☐   =   ☐

     198^2 = ☐   (remember to put a leading 'zero' in the result of step b.)

4) $187^2 =$

a) 187 − ☐ = ☐   X 2 = ☐

b) ☐ X ☐ = ☐

$187^2 =$ ☐
(remember to carry the left digit from b.)

---

5) $208^2 =$

a) 208 + ☐ = ☐   X 2 = ☐
$\quad\quad\quad$ (208 − 200)

b) ☐ X ☐ = ☐
$\;$ (208 − 200) $\;$ (208 − 200)

$208^2 =$ ☐
Put the result of a) and then the result of b).

---

6) $206^2 =$

a) $206^2 +$ ☐ = ☐   X 2 = ☐
$\quad\quad\quad$ (206 − 200)

b) ☐ X ☐ = ☐
$\;$ (206 − 200) $\;$ (206 − 200)

$206^2 =$ ☐

7) $202^2 =$

a) 202 + [ ] = [ ] X 2 = [ ]

b) [ ] X [ ] = [ ]

$202^2 =$ [ ]
(remember to put a leading 'zero' in the result of step b.)

---

8) $213^2 =$

a) 213 + [ ] = [ ] X 2 = [ ]

b) [ ] X [ ] = [ ]

$213^2 =$ [ ]
(remember to carry the digits.)

Exercises:

For the following, write down the answer directly:

1) 197^2 = ☐

2) 195^2 = ☐

3) 193^2 = ☐

4) 189^2 = ☐

5) 203^2 = ☐

6) 193^2 = ☐

7) 191^2 = ☐

8) 207^2 = ☐

9) 215^2 = ☐

10) 186^2 = ☐

11) 209^2 = ☐

12) 189^2 = ☐

13) 211^2 = ☐

Shortcut 41: Squaring numbers near 500

Rule
These are almost exactly same as squaring numbers near 200, just that you multiply the result of step a) by 5 instead of by 2.

Calculate the difference from 500 of the number to be squared.

If the number to be squared is less than 500, subtract the difference from it.
If the number to be squared is more than 500, add the difference to it.

Now, multiply this answer by 5.

This will form the left side digits of the answer.

Square the difference from 500 and put that as the two right side digits of the answer. As before, if the square is a single digit answer, put a '0' (zero) before it and if it is a three digit answer, take the right two digits and carry the third digit to the left side.

Illustration

Calculate 494^2
500 − 494 = 6
Since 494 is less than 500, subtract 6 from 494:
494 − 6 = 488

488 x 5 = 2440 (this is the added step compared to squaring numbers near 100.)

6^2 = 36

494^2 = 244036

Calculate 499^2
500 − 499 = 1
499 − 1 = 498

498 x 5 = 2490

1^2 = 1 (since it is single digit, we put a leading 'zero' before the 1.)

499^2 = 249001

Examples:
Calculate 488²
500 − 488 = 12
488 − 12 = 476

476 X 5 = 2380

12² = 144 (we carry the digit '1' to the left side.)

488² = 2380₁44

488² = 238144

Calculate 504²
504 − 500 = 4
Since 504 is more than 500, add 4 to 504.
504 + 4 = 508

508 X 5 = 2540

4² = 16

504² = 254016

Calculate 501²
501 − 500 = 1
501 + 1 = 502

502 X 5 = 2510

1² = 1 (we put a leading 'zero' before the 1.)

501² = 251001

Calculate 512²
512 − 500 = 12
512 + 12 = 524
524 X 5 = 2620

12² = 144 (we carry the digit '1' to the left side.)

512² = 2620₁44

512² = 262144

Exercise:

1) $492^2 =$

a) 492 − ☐ = ☐ × 5 = ☐
      (500 − 492)   (times 5)

b) ☐ × ☐ = ☐
   (500 − 492) (500 − 492)

$492^2 =$ ☐
Put the result of a) and then the result of b).

---

2) $496^2 =$

a) 496 − ☐ = ☐ × 5 = ☐
      (500 − 496)

b) ☐ × ☐ = ☐
   (500 − 496) (500 − 496)

$496^2 =$ ☐

---

3) $498^2 =$

a) 498 − ☐ = ☐ × 5 = ☐
      (500 − 498)

b) ☐ × ☐ = ☐
   (500 − 498) (500 − 498)

$498^2 =$ ☐   (remember to put a leading 'zero' in the result of step b.)

4) $487^2 =$

a) 487 − ☐ = ☐ X 5 = ☐
          (500 − 487)

b) ☐   X ☐   = ☐
    (500 − 487)   (500 − 487)

  $487^2 =$ ☐
  (remember to carry the left digit from b.)

---

5) $508^2 =$

a) 508 + ☐ = ☐ X 5 = ☐
          (508 − 500)

b) ☐   X ☐   = ☐
    (508 − 500)   (508 − 500)

  $508^2 =$ ☐
  Put the result of a) and then the result of b).

---

6) $506^2 =$

a) 506 + ☐ = ☐ X 5 = ☐
          (506 − 500)

b) ☐   X ☐   = ☐
    (506 − 500)   (506 − 500)

  $506^2 =$ ☐

7) $502^2 =$

a) 502 + ☐ = ☐ X 5 = ☐

b) ☐ X ☐ = ☐

$502^2 =$ ☐
(remember to put a leading 'zero' in the result of step b.)

8) $513^2 =$

a) 513 + ☐ = ☐ X 5 = ☐

b) ☐ X ☐ = ☐

$513^2 =$ ☐
(remember to carry the digits.)

Exercises:

For the following, write down the answer directly:

1) 497^2 = ☐

2) 495^2 = ☐

3) 493^2 = ☐

4) 489^2 = ☐

5) 503^2 = ☐

6) 493^2 = ☐

7) 491^2 = ☐

8) 507^2 = ☐

9) 515^2 = ☐

10) 486^2 = ☐

11) 509^2 = ☐

12) 489^2 = ☐

13) 511^2 = ☐

Shortcut 42: Squaring numbers that end in 25

Rule
1) Look at the part of the number other than 5 and add 3 to it.
2) Take the part other than 25 and multiply it with the result in step 1.
3) Put 625 after the result in step 2.

Illustration

Calculate 125^2
If we ignore the 5, we have 12, add 3 to it.
12 + 3 = 15

Take the part other than 25 and that is 1, multiply it with 15 to get:
15 X 1 = 15

Put 625 after this to get:
125^2 = 15625

Examples:
Calculate 525^2
Ignore the 5, we have 52
52 + 3 = 55

Ignore the 25 and we have 5:
55 X 5 = 275

Put 625 after this to get:
525^2 = 275625

Calculate 725^2
Ignore the 5, we have 72
72 + 3 = 75

Ignore the 25 and we have 7:
75 X 7 = 525

Put 625 after this to get:
725^2 = 525625

Exercise:

1) $225^2 =$

   ☐ + 3 = ☐
   (part of the number other than the 5)

   ☐ × ☐ = ☐
   (number above)   (part of number other than 25)

   $225^2 =$ ☐
   (Put 625 after the result you get above.)

---

2) $325^2 =$

   ☐ + 3 = ☐
   (part of the number other than the 5)

   ☐ × ☐ = ☐
   (number above)   (part of number other than 25)

   $325^2 =$ ☐
   (Put 625 after the result you get above.)

---

3) $425^2 =$

   ☐ + 3 = ☐

   ☐ × ☐ = ☐

   $425^2 =$ ☐
   (Put 625 after the result you get above.)

4) $625^2 =$

   [ ] + 3 = [ ]

   [ ] × [ ] = [ ]

   $625^2 =$ [ ]
   (Put 625 after the result you get above)

---

5) $825^2 =$

   [ ] + 3 = [ ]

   [ ] × [ ] = [ ]

   $825^2 =$ [ ]

---

6) $925^2 =$

   [ ] + 3 = [ ]

   [ ] × [ ] = [ ]

   $925^2 =$ [ ]

7) $1025^2 =$

   ☐ + 3 = ☐

   ☐ × ☐ = ☐

   $1025^2 =$ ☐

---

8) $2025^2 =$

   ☐ + 3 = ☐

   ☐ × ☐ = ☐

   $2025^2 =$ ☐

---

9) $4025^2 =$

   ☐ + 3 = ☐

   ☐ × ☐ = ☐

   $4025^2 =$ ☐

Shortcut 43: Squaring numbers in the 300s

Rule
1. Square the right most two digits. Keep the right side two digits of the answer and carry the rest to next step.
2. Multiply the right side two digits by 6 and add the carry from the previous step. Of this answer, keep the right side two digits and carry the rest.
3. Add the carry of the previous step to 9.
4. Put together the results of step 3, 2 and 1 from left to right.

Illustration

Calculate 321^2
1) $21^2 = 441$ (keep the 41 and carry 4)   41
2) $6 \times 21 = 126 + 4$ (carry) $= 130$. Keep 30 and carry 1   30
3) $1 + 9 = 10$ 10

$321^2 = 103041$

Examples:

Calculate 311^2
1) $11^2 = 121$ (keep the 21 and carry 1)   21
2) $6 \times 11 = 66 + 1$ (carry) $= 67$. Keep 67, no carry   67
3) $0 + 9 = 9$ 9

$311^2 = 96721$

Calculate 339^2
1) $39^2 = 1521$ (keep the 21 and carry 15)   21
2) $6 \times 39 = 234 + 15$ (carry) $= 249$. Keep 49 and carry 2   49
3) $2 + 9 = 11$ 11

$339^2 = 114921$

Exercise:

1) $308^2 =$

   $08^2 =$ ☐ Square right side 2 digits   ☐ Keep right 2 digits and carry the rest

   $6 \times 08 =$ ☐ + ☐ 6 X the right side 2 digits + carry   ☐ Keep right 2 digits and carry the rest

   carry + 9 =   ☐

   $308^2 =$ ☐

---

2) $322^2 =$

   $22^2 =$ ☐ Square right side 2 digits   ☐ Keep right 2 digits and carry the rest

   $6 \times 22 =$ ☐ + ☐ 6 X the right side 2 digits + carry   ☐ Keep right 2 digits and carry the rest

   carry + 9 =   ☐

   $322^2 =$ ☐

---

3) $329^2 =$

   $29^2 =$ ☐ Square right side 2 digits   ☐ Keep right 2 digits and carry the rest

   $6 \times 29 =$ ☐ + ☐ 6 X the right side 2 digits + carry   ☐ Keep right 2 digits and carry the rest

   carry + 9 =   ☐

   $329^2 =$ ☐

4) 317^2 =

17^2 = [ ]   [ ]
      Square right side 2 digits   Keep right 2 digits and carry the rest

6 X 17 = [ ]   [ ]
      6 times + carry. Keep 2 digits   Keep right 2 digits and carry the rest

carry + 9 =   [ ]

317^2 = [ ]

Exercises:
For the questions below, write down the 3 parts of the answer directly:

Step 3 (carry + 9)	Step 2 (6 times the two right digits + carry)	Step 1 (Square of the two right digits)

322^2 =

319^2 =

332^2 =

345^2 =

361^2 =

375^2 =

391^2 =

Shortcut 44: Squaring numbers in the 400s

Rule
1. Square the right most two digits. Keep the right side two digits of the answer and carry the rest to next step.
2. Multiply the right side two digits by 8 and add the carry from the previous step. Of this answer, keep the right side two digits and carry the rest.
3. Add the carry of the previous step to 16.
4. Put together the results of step 3, 2 and 1 from left to right.

Illustration

Calculate 415^2
1) $15^2 = 225$ (keep the 25 and carry 2)   25
2) $8 \times 15 = 120 + 2$ (carry) $= 122$. Keep 22 and carry 1   22
3) $1 + 16 = 17$ 17

$415^2 = 172225$

Examples:
Calculate 411^2
1) $11^2 = 121$ (keep the 21 and carry 1)   21
2) $8 \times 11 = 88 + 1$ (carry) $= 89$. Keep 89, no carry   89
3) $0 + 16 = 16$ 16

$411^2 = 168921$

Calculate 439^2
1) $39^2 = 1521$ (keep the 21 and carry 15)   21
2) $8 \times 39 = 312 + 15$ (carry) $= 327$. Keep 27 and carry 3   27
3) $3 + 16 = 19$ 19

$439^2 = 192721$

Exercise:

1) $408^2 =$

   $08^2 =$ ☐   ☐
   Square right side 2 digits   Keep right 2 digits and carry the rest

   $8 \times 08 =$ ☐ +   ☐
   8 X the right side 2 digits + carry   Keep right 2 digits and carry the rest

   carry + 16 =   ☐

   $408^2 =$ ☐

---

2) $422^2 =$

   $22^2 =$ ☐   ☐
   Square right side 2 digits   Keep right 2 digits and carry the rest

   $8 \times 22 =$ +   ☐
   8 X the right side 2 digits + carry   Keep right 2 digits and carry the rest

   carry + 16 =   ☐

   $422^2 =$ ☐

---

3) $429^2 =$

   $29^2 =$ ☐   ☐
   Square right side 2 digits   Keep right 2 digits and carry the rest

   $8 \times 29 =$ +   ☐
   8 X the right side 2 digits + carry   Keep right 2 digits and carry the rest

   carry + 16 =   ☐

   $429^2 =$ ☐

4) 417^2 =

17^2 = ☐
Square and keep 2 digits

8 X 17 = ☐
8 times + carry. Keep 2 digits

carry + 16 = ☐

417^2 = ☐

Exercises:
For the questions below, write down the 3 parts of the answer directly:

	Step 3 (carry + 16)	Step 2 (8 times the two right digits + carry)	Step 1 (square of the right two digits)

412^2 =

419^2 =

432^2 =

445^2 =

461^2 =

475^2 =

491^2 =

Shortcut 45: Sum of squares of first 'n' numbers

Rule
$1^2 + 2^2 + 3^2 + ... + (n-1)^2 + n^2 =$
To calculate the square of first 'n' numbers, find the value of:

$$\frac{(n) \times (n+1) \times (2n+1)}{6}$$

Illustration

$1^2 + 2^2 + 3^2 + 4^2 + 5^2 =$

Here n = 5 because we want to find the sum of square of the first five numbers.

n = 5
n + 1 = 6
2n + 1 = 11

$1^2 + 2^2 + 3^2 + 4^2 + 5^2 = \dfrac{5 \times 6 \times 11}{6}$

= 5 x 11
= 55

$1^2 + 2^2 + ... + 9^2 + 10^2 =$

n = 10
n + 1 = 11
2n + 1 = 21

$1^2 + 2^2 + ... + 9^2 + 10^2 = \dfrac{10 \times 11 \times 21}{6}$

= 385

Exercise:

1) $1^2 + 2^2 + \ldots + 14^2 + 15^2 =$

   $n =$ ☐
   $n + 1 =$ ☐
   $2n + 1 =$ ☐

   $$= \frac{1^2 + 2^2 + \ldots + 14^2 + 15^2}{6} = \frac{\boxed{} \times \boxed{} \times \boxed{}}{6}$$

   $=$ ☐

2) $1^2 + 2^2 + \ldots + 29^2 + 30^2 =$

   $n =$ ☐
   $n + 1 =$ ☐
   $2n + 1 =$ ☐

   $$= \frac{1^2 + 2^2 + 3^2 + \ldots + 29^2 + 30^2}{6} = \frac{\boxed{} \times \boxed{} \times \boxed{}}{6}$$

   $=$ ☐

3) $1^2 + 2^2 + ... + 22^2 + 23^2 =$

  n = ☐
  n + 1 = ☐
  2n + 1 = ☐

$= \dfrac{1^2 + 2^2 + ... + 22^2 + 23^2}{6} = \dfrac{\boxed{} \times \boxed{} \times \boxed{}}{6}$

= ☐

---

4) $1^2 + 2^2 + ... + 51^2 + 52^2 =$

  n = ☐
  n + 1 = ☐
  2n + 1 = ☐

$= \dfrac{1^2 + 2^2 + ... + 51^2 + 52^2}{6} = \dfrac{\boxed{} \times \boxed{} \times \boxed{}}{6}$

= ☐

5) $1^2 + 2^2 + \ldots + 74^2 + 75^2 =$

$= \dfrac{\boxed{} \times \boxed{} \times \boxed{}}{6}$

6) $1^2 + 2^2 + \ldots + 99^2 + 100^2 =$

$= \dfrac{\boxed{} \times \boxed{} \times \boxed{}}{6}$

7) $1^2 + 2^2 + \ldots + 499^2 + 500^2 =$

$= \dfrac{\boxed{} \times \boxed{} \times \boxed{}}{6}$

8) $1^2 + 2^2 + \ldots + 999^2 + 1000^2 =$

$= \dfrac{\boxed{} \times \boxed{} \times \boxed{}}{6}$

Shortcut 46: Sum of squares of 'n' consecutive numbers

Example: What is $4^2 + 5^2 + 6^2 + 7^2 + 8^2 = ?$

Rule
Think of this as sum of squares of numbers from 'm' to 'n'.
$m^2 + ... + n^2$

We will use the previous shortcut to solve this problem by thinking of $4^2 + 5^2 + 6^2 + 7^2 + 8^2$ as:

$(1^2 + 2^2 + ... + 7^2 + 8^2) - (1^2 + 2^2 + 3^2)$

This is because we know how to calculate the sum of squares of first 'n' numbers.

In general terms: to calculate the sum of squares of consecutive numbers from 'm' to 'n', we find the sum of squares of first 'n' integers and from that subtract the sum of squares from 1 to (m – 1).

Illustration

$4^2 + 5^2 + 6^2 + 7^2 + 8^2 =$
$(1^2 + 2^2 + ... + 7^2 + 8^2) - (1^2 + 2^2 + 3^2)$

$(1^2 + 2^2 + ... + 7^2 + 8^2)$
n = 8
n + 1 = 9
2n + 1 = 17

$= \dfrac{8 \times 9 \times 17}{6} = 204$

$(1^2 + 2^2 + 3^2)$
n = 3
n + 1 = 4
2n + 1 = 7

$= \dfrac{3 \times 4 \times 7}{6} = 14$

So: $4^2 + 5^2 + 6^2 + 7^2 + 8^2 = 204 - 14 = 190$.

Exercise:

1) $5^2 + 6^2 + \ldots + 19^2 + 20^2 =$

$= (1^2 + 2^2 + \ldots + 19^2 + 20^2) - (1^2 + 2^2 + 3^2 + 4^2) =$

$= \dfrac{\boxed{} \times \boxed{} \times \boxed{}}{6} - \dfrac{\boxed{} \times \boxed{} \times \boxed{}}{6}$

$= \boxed{} - \boxed{}$

$= \boxed{}$

---

2) $7^2 + 8^2 + \ldots + 29^2 + 30^2 =$

$= (1^2 + 2^2 + \ldots + 29^2 + 30^2) - (1^2 + 2^2 + \ldots + 5^2 + 6^2) =$

$= \dfrac{\boxed{} \times \boxed{} \times \boxed{}}{6} - \dfrac{\boxed{} \times \boxed{} \times \boxed{}}{6}$

$= \boxed{} - \boxed{}$

$= \boxed{}$

---

3) $11^2 + 12^2 + \ldots + 49^2 + 50^2 =$

$= (1^2 + 2^2 + \ldots + 49^2 + 50^2) - (1^2 + 2^2 + \ldots + 9^2 + 10^2) =$

$= \dfrac{\boxed{} \times \boxed{} \times \boxed{}}{6} - \dfrac{\boxed{} \times \boxed{} \times \boxed{}}{6}$

$= \boxed{} - \boxed{}$

$= \boxed{}$

4) $26^2 + 27^2 + \ldots + 49^2 + 50^2 =$

$= (1^2 + 2^2 + \ldots + 49^2 + 50^2) - (1^2 + 2^2 + \ldots + 24^2 + 25^2) =$

$= \dfrac{\boxed{} \times \boxed{} \times \boxed{}}{6} - \dfrac{\boxed{} \times \boxed{} \times \boxed{}}{6}$

$= \boxed{} - \boxed{}$

$= \boxed{}$

---

5) $101^2 + 102^2 + \ldots + 199^2 + 200^2 =$

$= (1^2 + 2^2 + \ldots + 199^2 + 200^2) - (1^2 + 2^2 + \ldots + 99^2 + 100^2) =$

$= \dfrac{\boxed{} \times \boxed{} \times \boxed{}}{6} - \dfrac{\boxed{} \times \boxed{} \times \boxed{}}{6}$

$= \boxed{} - \boxed{}$

$= \boxed{}$

---

6) $501^2 + 502^2 + \ldots + 999^2 + 1000^2 =$

$= (1^2 + 2^2 + \ldots + 999^2 + 1000^2) - (1^2 + 2^2 + \ldots + 499^2 + 500^2) =$

$= \dfrac{\boxed{} \times \boxed{} \times \boxed{}}{6} - \dfrac{\boxed{} \times \boxed{} \times \boxed{}}{6}$

$= \boxed{} - \boxed{}$

$= \boxed{}$

Shortcut 47: Sum of cubes of 'n' consecutive numbers

Rule
$1^3 + 2^3 + ... + (n-1)^3 + n^3 = ?$
To calculate the sum of cube of first 'n' numbers, find the value of:

$\left(\dfrac{(n) \times (n+1)}{2}\right)^2$ or written as $\dfrac{(n) \times (n+1)}{2} \times \dfrac{(n) \times (n+1)}{2}$

Illustration

$1^3 + 2^3 + 3^3 + 4^3 =$
Here n = 4 because we want to find the sum of cubes of the first four numbers.
n = 4
n + 1 = 5

$1^3 + 2^3 + 3^3 + 4^3 =$
$= \dfrac{(n) \times (n+1)}{2} \times \dfrac{(n) \times (n+1)}{2}$

$= \dfrac{(4 \times 5)}{2} \times \dfrac{(4 \times 5)}{2}$

$= 10 \times 10 = 100$

Examples:
$1^3 + 2^3 + ... + 9^3 + 10^3 =$
n = 10
n + 1 = 11

$1^3 + 2^3 + ... 9^3 + 10^3 =$
$= \dfrac{(10 \times 11)}{2} \times \dfrac{(10 \times 11)}{2}$

$= 55 \times 55$

$= 3025$

Exercise:

1) $1^3 + 2^3 + ... + 13^3 + 14^3 =$

   n   = ☐
   n + 1 = ☐

   $1^3 + 2^3 + ... + 13^3 + 14^3 =$

   $= \dfrac{\boxed{} \times \boxed{}}{2} \times \dfrac{\boxed{} \times \boxed{}}{2}$

   $= \boxed{} \times \boxed{}$

   $= \boxed{}$

2) $1^3 + 2^3 + ... + 19^3 + 20^3 =$

   n   = ☐
   n + 1 = ☐

   $1^3 + 2^3 + ... + 19^3 + 20^3 =$

   $= \dfrac{\boxed{} \times \boxed{}}{2} \times \dfrac{\boxed{} \times \boxed{}}{2}$

   $= \boxed{} \times \boxed{}$

   $= \boxed{}$

3) $1^3 + 2^3 + ... + 99^3 + 100^3 =$

n = ☐
n + 1 = ☐

$1^3 + 2^3 + ... + 99^3 + 100^3 =$

= $\dfrac{\boxed{} \times \boxed{}}{2}$ × $\dfrac{\boxed{} \times \boxed{}}{2}$

= ☐ × ☐

= ☐

4) $1^3 + 2^3 + ... + 199^3 + 200^3 =$

n = ☐
n + 1 = ☐

$1^3 + 2^3 + ... + 199^3 + 200^3 =$

= $\dfrac{\boxed{} \times \boxed{}}{2}$ × $\dfrac{\boxed{} \times \boxed{}}{2}$

= ☐ × ☐

= ☐

Shortcut 48: Adding Time

Rule
Add the hours and minutes of the time to be added separately.

If the minutes add to less than 60, then that is the answer.

But,

If the minutes add to more than 60, add 40 to it. Keep the right two digits as the minutes and add 1 (left digit) to the hours to get the answer.

Illustration

What is 6 hour 45 minutes + 8 hour 35 minutes?

Add the hours and minutes separately:
6 + 8 = 14 hours
45 + 35 = 80 minutes

Since the minutes part adds up to more than 60, we add 40 to it:
80 + 40 = 120

The right side two digits are the minutes and the left side '1' should be carried over to the hours to get:
14 + 1 = 15

So 6 hours 45 minutes + 8 hours 35 minutes = 15 hours and 20 minutes.

3 hours 55 minutes + 2 hours 50 minutes =

Add the hours and minutes separately:
3 + 2 = 5 hours
55 + 50 = 105 minutes.
Since it is more than 60, we add 40 to it:
105 + 40 = 145

We keep the right side two digits (45) and add the '1' to the number of hours:
5 + 1 = 6

So 3 hours and 55 minutes + 2 hours and 50 minutes = 6 hours and 45 minutes

Exercise:

1) 6 hours 45 minutes + 2 hours 50 minutes =

   Add the hours

   ☐ + ☐ = ☐

   Add the minutes

   ☐ + ☐ = ☐   If minutes are more than 60, add 40:   ☐ + 40 = ☐

   Keep the right two digits as minutes and add the '1' to the hours to get the answer.

   6 hours 45 minutes + 2 hours 50 minutes = ☐

2) 4 hours 32 minutes + 2 hours 48 minutes =

   Add the hours

   ☐ + ☐ = ☐

   Add the minutes

   ☐ + ☐ = ☐   If minutes are more than 60, add 40:   ☐ + 40 = ☐

   Keep the right two digits as minutes and add the '1' to the hours to get the answer.

   4 hours 32 minutes + 2 hours 48 minutes = ☐

3) 1 hours 12 minutes + 2 hours 38 minutes =

   Add the hours

   ☐ + ☐ = ☐

   Add the minutes

   ☐ + ☐ = ☐   If minutes are more than 60, add 40:   ☐ + 40 = ☐

   Keep the right two digits as minutes and add the '1' to the hours to get the answer.

   1 hours 12 minutes + 2 hours 38 minutes = ☐
   (Note: Here the minutes don't add up to more than 60. So you <u>should not add 40</u>).

4) 10 hours 27 minutes + 12 hours 56 minutes =

Add the hours

☐ + ☐ = ☐

Add the minutes

☐ + ☐ = ☐ + 40 = ☐

10 hours 27 minutes + 12 hours 56 minutes = ☐

---

5) 5 hours 45 minutes + 20 hours 30 minutes =

Add the hours

☐ + ☐ = ☐

Add the minutes

☐ + ☐ = ☐ + 40 = ☐

5 hours 45 minutes + 20 hours 30 minutes = ☐

---

6) 9 hours 25 minutes + 8 hours 15 minutes =

Add the hours

☐ + ☐ = ☐

Add the minutes

☐ + ☐ = ☐ + 40 = ☐

9 hours 25 minutes + 8 hours 15 minutes = ☐

7) 100 hours 57 minutes + 90 hours 56 minutes =

Add the hours

☐ + ☐ = ☐

Add the minutes

☐ + ☐ = ☐ + 40 = ☐

100 hours 57 minutes + 90 hours 56 minutes = ☐

Shortcut 49: Checking Answers: Digit Sums

Before we do the rule, we will first understand a concept called 'Digit Sum'.

Every number can be reduced to a single digit, called it's digit sum.
To get the digit sum, you simply add up all of the digits in the given number. If this sum is larger than one digit, you add the digits of this sum together too. You keep doing this until the result is a single digit. That is the digit sum of the original number.

Take 26 for example. 2 + 6 = 8, so the digit sum of 26 is 8.
Look at 296. 2 + 9 + 6 = 17. This is more than 9, so we can reduce it further so we add the 1 + 7 and get 8. So the digit sum of 296 is 8.
Now we look at 2345. 2 + 3 + 4 + 5 = 14.
1 + 4 = 5. Digit sum of 2345 is 5.

By definition: Digit sum of 9 = Digit sum of 0.

There are two important properties of digit sum that are very useful:
1. The digit sum does not change if you add or subtract 9 from the number, add or subtract any multiple of 9, or even insert / delete a set of 9's in it / from it.

Example.
   The digit sum of 26 is 8 (done above).
   - If we add 9 to 26, we get: 26 + 9 = 35. The digit sum of 35 = 3 + 5 = 8.
   - If we subtract 9 from 26, we get 26 - 9 = 17. The digit sum of 17 = 1 + 7 = 8.
   - If we add 9 x 15 (135) to 26, we get 26 + 135 = 161.
     Digit sum of 161 = 1 + 6 + 1 = 8.
   - If we insert a 9 in 26, we get 296.
     The digit sum of 296 = 2 + 9 + 6 = 17 = 1 + 7 = 8
   - If we insert a series of 9 in 26, we get (say) 29996.
     The digit sum of 29996 = 2 + 9 + 9 + 9 + 6 = 35 = 3 + 5 = 8.

2. As you perform mathematical operations on a set of numbers, the digit sums also change as if the operation has been performed on them.

So, if a X b = c, then digit sum (a) X digit sum (b) = digit sum (c).

Lets understand with an example:
   74 X 6 = 444.
   The property would mean that:
   = [(digit sum of 74) X (digit sum of 6)] = digit sum of (74 X 6)]
   Digit sum of:
   - 74 = 7 + 4 = 11 = 2
   - 6 = 6
   - 444 = 4 + 4 + 4 = 12 = 3

   digit sum (2 X 6) = digit sum (12) = 1 + 2 = 3 = digit sum of 444 !!!

While this does not help in getting the right answer, it has a brilliant application in spotting wrong answers.

Rule
In any mathematical calculation, calculate the digit sum of the numbers on which the calculation is being performed.

Perform the same calculation on the digit sums and the result should be equal to the digit sum of the answer of the original calculation.

If it does not match, then your answer of the original calculation is wrong.

786 + 687 = 1473

Digit Sum (786) = 7 + 8 + 6 = 21 = 2 + 1 = 3
Digit Sum (687) = 6 + 8 + 7 = 21 = 2 + 1 = 3

Since our original calculation is addition, we should add the digit sums of the two numbers and that should be equal to the digit sum of the answer.

Digit sum (786) + Digit sum (687) = 3 + 3 = 6
Digit sum (1473) = 1 + 4 + 7 + 3 = 15 = 1 + 5 = 6

It matches. What this means is that our answer is possibly correct. If it did not match, then we can say it is definitely wrong.

How do we understand this?

Suppose, instead of 1473, we got the incorrect answer of 1437 (3 and 7 are interchanged).
In this case also, the digit sum will be 1 + 4 + 3 + 7 = 15 = 1 + 5 = 6. Because the interchange of digits does not affect the digit sum, we should not assume that because the digit sum matches, the answer must be correct.

The reverse however is always true. If the digit sum does not match, the answer is definitely wrong.

Illustration: Lets do a few more examples to establish that this works.

542 x 642 = 347964

Digit sum (542) = 5 + 4 + 2 = 11 = 1 + 1 = 2
Digit sum (642) = 6 + 4 + 2 = 12 = 1 + 2 = 3

Digit sum (542 X 642) = digit sum (542) X digit sum (642) = 2 X 3 = 6

Digit sum (347964) = 3 + 4 + 7 + 9 + 6 + 4 = 33 = 3 + 3 = 6

Lets now do an example of an incorrect calculation:

345 X 6783 = 2346135

Digit sum (345) = 3 + 4 + 5 = 12 = 1 + 2 = 3
Digit sum (6783) = 6 + 7 + 8 + 3 = 24 = 2 + 4 = 6

Digit sum (345 X 6783) = Digit sum (345) X Digit sum (6783)
= 3 X 6 = 18 = 1 + 8 = 9

Digit sum(2346135) = 2 + 3 + 4 + 6 + 1 + 3 + 5 = 24 = 2 + 4 = 6

This does not match, so definitely the answer is incorrect!

In summary, if you want to ensure you don't have an incorrect answer (versus ensuring that you have a correct answer), you can use the digit sum of the numbers before and after the mathematical operation.

Exercise: Calculate the digit sum of the following numbers

Digit sum (72) =

Digit sum (6357) =

Digit sum (346677) =

Digit sum (675) =

Digit sum (566556) =

Digit sum (436436526) =

Digit sum (43556) =

Digit sum (132443) =

Digit sum (239987987) =

Digit sum (9768786876698) =

<u>Are the following multiplications 'Definitely Incorrect' or 'Possibly Correct'. Circle the correct one.</u>

23 X 76 =

Digit sum (23) X Digit sum (76) =
Digit sum (1748) =

Digit sums don't match?
Digit sums match?

1748

Definitely Incorrect
Possibly Correct

873 X 246 =

Digit sum (873) X Digit sum (246) =
Digit sum (215758) =

Digit sums don't match?
Digit sums match?

215758

Definitely Incorrect
Possibly Correct

432 X 72 =   31204

Digit sum (432) X Digit sum (72) =
Digit sum (31204) =

Digit sums don't match?   Definitely Incorrect
Digit sums match?   Possibly Correct

712 X 722 =   514064

Digit sum (712) X Digit sum (722) =
Digit sum (514064) =

Digit sums don't match?   Definitely Incorrect
Digit sums match?   Possibly Correct

Shortcut 50: Cubing 2-digit Numbers

Rule
To begin with, you must know the cubes of the first 10 numbers:

$1^3 = 1$ $2^3 = 8$ $3^3 = 27$ $4^3 = 64$ $5^3 = 125$

$6^3 = 216$ $7^3 = 343$ $8^3 = 512$ $9^3 = 729$ $10^3 = 1000$

Let's understand this rule through an example:
$14^3 =$

First thing we should calculate is the ratio of the tens digit and the units digit. We will call this R. In this example, R = ¼.

The answer will have 4 sections (lets call them a, b, c and d). We will write down the parts of the answer from the right (starting with d).

1. d: Find the cube of the units place digit: $4^3 = 64$. We will use 64 in the next step also. Keep 4 as part of 'd' and carry 6 to the next step.

2. c: Multiply the 64 we got above by R, multiply by 3 and to it add the carry from previous step. Keep the right most digit there and carry the rest to the next number on the left.
So we do 64 X ¼ = 16 X 3 = 48 + 6 (carry over) = 54.
We will use 16 in the next step also.
Keep 4 and carry 5.

3. b: Multiply the 16 in the above step by R, multiply by 3 and add the carry:
16 X ¼ = 4 X 3 = 12 + 5 (carry over) = 17.
Keep the 7 and carry the 1. We will use the 4 in the next step also.

4. a: Multiply the 4 by R and add the carry: 4 X ¼ = 1 + 1 (carry over) = 2.

Now write a, b, c and d in order to get the answer:

$14^3 = 2744$

The steps above seem complicated. As you do a few more examples, you will see it in very simple to understand and to use.

The above explanation looks a little complicated. So we will summarize the rule here and do one more example to understand in more detail the steps followed:

Rule

	Step		Keep
1.	Calculate the ratio of tens and units digit = R.		
d.	Cube the units digit.		Right digit and carry the rest.
c.	Multiply the number above by R.	Multiply by 3 and add carry over.	Right digit and carry the rest.
b.	Multiply the number above by R.	Multiply by 3 and add carry over.	Right digit and carry the rest.
a.	Multiply the number above by R.	Add carry over.	All the digits.
	Answer is 'a', 'b', 'c', and 'd' written in order.		

Example:

Calculate 16^3

	Step		Keep
1.	$R = \frac{1}{6}$		
d.	$6 \times 6 \times 6 = 216$		d = 6 Carry = 21
c.	$216 \times \frac{1}{6} = 36$	$36 \times 3 = 108 + 21$ $= 129$	c = 9 Carry = 12
b.	$36 \times \frac{1}{6} = 6$	$6 \times 3 = 18$ $18 + 12 = 30$	b = 0 Carry = 3
a.	$6 \times \frac{1}{6} = 1$	$1 + 3 = 4$	a = 4
	16^3		4096

Example:

Calculate 25³

	Step		Keep
1.	$R = \frac{2}{5}$		
d.	5 X 5 X 5 = 125		d = 5 Carry = 12
c.	125 X $\frac{2}{5}$ = 50	50 X 3 = 150 + 12 = 162	c = 2 Carry = 16
b.	50 X $\frac{2}{5}$ = 20	20 X 3 = 60 60 + 16 = 76	b = 6 Carry = 7
a.	20 X $\frac{2}{5}$ = 8	8 + 7 = 15	a = 15
	25³		15625

Calculate 48³

	Step		Keep
1.	$R = \frac{4}{8} = \frac{1}{2}$		
d.	8 X 8 X 8 = 512		d = 2 Carry = 51
c.	512 X $\frac{1}{2}$ = 256	256 X 3 = 768 768 + 51 = 819	c = 9 Carry = 81
b.	256 X $\frac{1}{2}$ = 128	128 X 3 = 384 384 + 81 = 465	b = 5 Carry = 46
a.	128 X $\frac{1}{2}$ = 64	64 + 46 = 110	a = 110
	48³		110592

Exercise:

1) Calculate 12^3

	Step		Keep
1.	R =		
d.			d = Carry =
c.			c = Carry =
b.			b = Carry =
a.			a =
	12^3		

2) Calculate 21^3

	Step		Keep
1.	R =		
d.			d = Carry =
c.			c = Carry =
b.			b = Carry =
a.			a =
	21^3		

3) Calculate 33^3

	Step		Keep
1.	R =		
d.			d = Carry =
c.			c = Carry =
b.			b = Carry =
a.			a =
	33^3		

4) Calculate 62^3

	Step		Keep
1.	R =		
d.			d = Carry =
c.			c = Carry =
b.			b = Carry =
a.			a =
	62^3		

5) Calculate 86³

	Step		Keep
1.	R =		
d.			d = Carry =
c.			c = Carry =
b.			b = Carry =
a.			a =
	86³		

6) Calculate 99³

	Step		Keep
1.	R =		
d.			d = Carry =
c.			c = Carry =
b.			b = Carry =
a.			a =
	99³		

Made in the USA
San Bernardino, CA
03 April 2013